INNSKY AIR FRYER COOKBOOK

500

Crispy, Easy, Healthy, Fast & Fresh Recipes For Your Innsky Air Fryer (Recipe Book)

VANESSA BRIGHT

Copyright

Table Of Contents

INTRODUCTION

The Innsky Air fryer is an easy way to cook delicious healthy meals. Rather than cooking the food in oil and hot fat that may affect your health, the machine uses rapid hot air to circulate around and cook meals. This allows the outside of your food to be crispy and ensures that the inside layers are cooked through.

The Innsky Air fryer allows us to cook almost everything and a lot of dishes. We can use the Innsky Air fryer for cooking Meat, vegetables, poultry, fruit, fish, and a wide variety of desserts. It is possible to prepare your entire meals, starting from appetizers to main courses and desserts. Not to mention, The Innsky Air fryer also allows homemade preserves or even delicious sweets and cakes.

How Does the Innsky Air fryer Works?

The technology of the Innsky Air fryer is very simple. Fried foods get their crunchy texture because hot oil heats foods quickly and evenly on their surface. Oil is an excellent heat conductor, which helps with fast and simultaneous cooking across all ingredients. For decades cooks have used convection ovens to mimic the effects of frying or cooking the whole surface of the food. But the air never circulates quickly enough to achieve that delicious surface crisp we all love in fried foods.

With this mechanism, the air is circulated on high degrees, up to 200° C, to "air fry" any food such as fish, chicken or chips, etc. This technology has changed the whole idea of cooking by reducing the fat up to 80% compared to old-fashioned deep fat frying.

The Innsky Air fryer cooking releases the heat through a heating element that cooks the food more healthily and appropriately. There's also an exhaust fan right above the cooking chamber, which provides the food required airflow. This way, food is cooked with constant heated air. This leads to the same heating temperature reaching every single part of the food that is being cooked. So, this is an only grill and the exhaust fan that is helping the Innsky Air fryer to boost air at a constantly high speed to cook healthy food with less fat.

The internal pressure increases the temperature that will then be controlled by the exhaust system. Exhaust fan also releases extra filtered air to cook the food in a much healthier way. The Innsky Air fryer has no odor at all, and it is absolutely harmless, making it user and environment-friendly.

Benefits of the Innsky Air Fryer:

- Healthier, oil-free meals
- It eliminates cooking odors through internal air filters
- Makes cleaning easier due to lack of oil grease
- The Innsky Air Fryer can bake, grill, roast and fry providing more options
- A safer method of cooking compared to deep frying with exposed hot oil
- Has the ability to set and leave, as most models and it includes a digital timer

The Innsky Air fryer is an all-in-one that allows cooking to be easy and quick. It also leads to a lot of possibilities once you get to know it. Once you learn the basics and become familiar with your Innsky Air fryer, you can feel free to experiment and modify the recipes in the way you prefer. You can prepare a vast number of dishes in the Innsky Air fryer, and you can adapt your favorite stove-top dish, so it becomes air fryer–friendly. It all boils down to variety and lots of options, right?

Cooking perfect and delicious as well as healthy meals has never been easier. You can see how this recipe collection proves itself.

Enjoy!

Breakfast
Air Fryer Hard-Boiled Eggs
PREP: 1 MINUTE • COOK TIME: 15 MINUTES • TOTAL: 16 MINUTES• SERVES: 6

Ingredients
6 eggs
Instructions
1   Place the eggs in the air fryer basket. (You can put the eggs in an oven-safe bowl if you are worried about them rolling around and breaking.)
2   Set the temperature of your Innsky AF to 250°F. Set the timer and bake for 15 minutes (if you prefer a soft-boiled egg, reduce the cook time to 10 minutes). Meanwhile, fill a medium mixing bowl half full of ice water. Use tongs to remove the eggs from the air fryer basket, and transfer them to the ice water bath. Let the eggs sit for 5 minutes in the ice water.
    Peel and eat on the spot or refrigerate for up to 1 week.
Per Serving: Calories: 72; Fat: 5g; Saturated fat: 2g; Carbohydrate: 0g; Fiber: 0g; Sugar: 0g; Protein: 6g; Iron: 1mg; Sodium: 70mg

Easy Air Fryer Baked Eggs with Cheese
PREP: 2 MINUTES • COOK TIME: 6 MINUTES • TOTAL: 8 MINUTES •SERVES: 2

Ingredients
2 large eggs
2 tablespoons half-and-half, divided
2 teaspoons shredded Cheddar cheese, divided

Salt
Freshly ground black pepper

Instructions
1   Lightly coat the insides of 2 (8-ounce) ramekins with cooking spray. Break an egg into each ramekin. Add 1 tablespoon of half-and-half and 1 teaspoon of cheese to each ramekin. Season with salt and pepper. Using a fork, stir the egg mixture. Set the ramekins in the air fryer basket.
2   Set the temperature of your Innsky AF to 330°F. Set the timer and bake for 6 minutes. Check the eggs to make sure they are cooked. If they are not done, cook for 1 minute more and check again.

Bacon-and-Eggs Avocado
PREP: 5 MINUTES • COOK TIME: 17 MINUTES • TOTAL: 22 MINUTES • SERVES: 1

Ingredients
1 large egg

1 avocado, halved, peeled, and pitted

2 slices bacon

Fresh parsley, for serving (optional)

Sea salt flakes, for garnish (optional)

Instructions
1   Spray the Innsky air fryer basket with avocado oil. Preheat the Innsky air fryer to 320°F. Fill a small bowl with cool water.

2   Soft-boil the egg: Place the egg in the air fryer basket. Cook for 6 minutes for a soft yolk or 7 minutes for a cooked yolk. Transfer the egg to the bowl of cool water and let sit for 2 minutes. Peel and set aside. Use a spoon to carve out extra space in the center of the avocado halves until the cavities are big enough to fit the soft-boiled egg. Place the soft-boiled egg in the center of one half of the avocado and replace the other half of the avocado on top, so the avocado appears whole on the outside.

Starting at one end of the avocado, wrap the bacon around the avocado to completely cover it. Use toothpicks to hold the bacon in place. Place the bacon-wrapped avocado in the Innsky air fryer basket and cook for 5 minutes. Flip the avocado over and cook for another 5 minutes, or until the bacon is cooked to your liking. Serve on a bed of fresh parsley, if desired, and sprinkle with salt flakes, if desired.

Best served fresh. Store extras in an airtight container in the fridge for up to 4 days. Reheat in a preheated 320°F air fryer for 4 minutes, or until heated through.

Per serving: Calories **536**; Fat **46g**; Protein **18g**; Total carbs **18g**; Fiber **14g**

Double-Dipped Mini Cinnamon Biscuits

PREP: 15 MINUTES • COOK TIME: 13 MINUTES • TOTAL: 28 MINUTES • SERVES: 8 BISCUITS

Ingredients

2 cups blanched almond flour

½ cup Swerve confectioners'-style sweetener or equivalent amount of liquid or powdered sweetener

1 teaspoon baking powder

½ teaspoon fine sea salt

¼ cup plus 2 tablespoons (¾ stick) very cold unsalted butter

¼ cup unsweetened, unflavored almond milk

1 large egg

1 teaspoon vanilla extract

3 teaspoons ground cinnamon

GLAZE:

½ cup Swerve confectioners'-style sweetener or equivalent amount of powdered sweetener

¼ cup heavy cream or unsweetened, unflavored almond milk

Instructions

1 Preheat the Innsky air fryer to 350°F. Line a pie pan that fits into your air fryer with parchment paper.
   In a medium-sized bowl, mix together the almond flour, sweetener (if powdered; do not add liquid sweetener), baking powder, and salt. Cut the butter into ½-inch squares, then use a hand mixer to work the butter into the dry ingredients. When you are done, the mixture should still have chunks of butter.
   In a small bowl, whisk together the almond milk, egg, and vanilla extract (if using liquid sweetener, add it as well) until blended. Using a fork, stir the wet ingredients into the dry ingredients until large clumps form. Add the cinnamon and use your hands to swirl it into the dough. Form the dough into sixteen 1-inch balls and place them on the prepared pan, spacing them about ½ inch apart. (If you're using a smaller air fryer, work in batches if necessary.)

2 Bake in the air fryer until golden, 10 to 13 minutes. Remove from the air fryer and let cool on the pan for at least 5 minutes.
   While the biscuits bake, make the glaze: Place the powdered sweetener in a small bowl and slowly stir in the heavy cream with a fork. When the biscuits have cooled somewhat, dip the tops into the glaze, allow it to dry a bit, and then dip again for a thick glaze. Serve warm or at room temperature. Store unglazed biscuits in an airtight container in the refrigerator for up to 3 days or in the freezer for up to a month. Reheat in a preheated 350°F air fryer for 5 minutes, or until warmed through, and dip in the glaze as instructed above.

Per serving: Calories 546; Fat 51g; Protein 14g; Total carbs 13g; Fiber 6g

Foolproof Air Fryer Bacon
PREP: 5 MINUTES • COOK TIME: 10 MINUTES • TOTAL: 15 MINUTES •SERVES: 5

Ingredients
10 slices bacon

Instructions

1 Cut the bacon slices in half, so they will fit in the air fryer.
   Place the half-slices in the fryer basket in a single layer. (You may need to cook the bacon in more than one batch.)

2 Set the temperature of your Innsky AF to 400°F. Set the timer and fry for 5 minutes. Open the drawer and check the bacon. (The power of the fan may have caused the bacon to fly around during the cooking process. If so, use a fork or tongs to rearrange the slices.)
   Reset the timer and fry for 5 minutes more. When the time has elapsed, check the bacon again. If you like your bacon crispier, cook it for another 1 to 2 minutes.

Per Serving: Calories: 87; Fat: 7g; Saturated fat: 2g; Carbohydrate: 0g; Fiber: 0g; Sugar: 0g; Protein: 6g; Iron: 0mg; Sodium: 370mg

Meritage Eggs
PREP: 5 MINUTES • COOK TIME: 8 MINUTES • TOTAL: 13 MINUTES • SERVES: 2

Ingredients

2 teaspoons unsalted butter (or coconut oil for dairy-free), for greasing the ramekins

4 large eggs

2 teaspoons chopped fresh thyme

½ teaspoon fine sea salt

¼ teaspoon ground black pepper

2 tablespoons heavy cream (or unsweetened, unflavored almond milk for dairy-free)

3 tablespoons finely grated Parmesan

Fresh thyme leaves, for garnish (optional)

Instructions

1 Preheat the Innsky air fryer to 400°F. Grease two 4-ounce ramekins with the butter. Crack 2 eggs into each ramekin and divide the thyme, salt, and pepper between the ramekins. Pour 1 tablespoon of the heavy cream into each ramekin. Sprinkle each ramekin with 1½ tablespoons of the Parmesan cheese.

2 Place the ramekins in the air fryer and cook for 8 minutes for soft-cooked yolks (longer if you desire a harder yolk). Garnish with a sprinkle of ground black pepper and thyme leaves, if desired. Best served fresh.

Per serving: Calories 331; Fat 29g; Protein 16g; Total carbs 2g; Fiber 0.2g

Easy Air Fryer Buttermilk Biscuits
PREP: 5 MINUTES • COOK TIME: 5 MINUTES • TOTAL: 10 MINUTES • SERVES: 12

Ingredients

2 cups all-purpose flour
1 tablespoon baking powder
¼ teaspoon baking soda
2 teaspoons sugar
1 teaspoon salt

6 tablespoons (¾ stick) cold unsalted butter, cut into 1-tablespoon slices
¾ cup buttermilk
4 tablespoons (½ stick) unsalted butter, melted (optional)

Instructions

1   Spray the Innsky air fryer basket with olive oil.
In a large mixing bowl, combine the flour, baking powder, baking soda, sugar, and salt and mix well.
Using a fork, cut in the butter until the mixture resembles coarse meal.
Add the buttermilk and mix until smooth. Sprinkle flour on a clean work surface. Turn the dough out onto the work surface and roll it out until it is about ½ inch thick.
Using a 2-inch biscuit cutter, cut out the biscuits. Place the uncooked biscuits in the greased Innsky air fryer basket in a single layer.

2   Set the temperature of your Innsky AF to 360°F. Set the timer and bake for 5 minutes.
Transfer the cooked biscuits from the air fryer to a platter. Brush the tops with melted butter, if desired. Cut the remaining biscuits (you may have to gather up the scraps of dough and reroll the dough for the last couple of biscuits). Bake the remaining biscuits.
Plate, serve, and enjoy!

Per Serving (1 biscuit): Calories: 146; Fat: 6g; Saturated fat: 4g; Carbohydrate: 20g; Fiber: 1g; Sugar: 2g; Protein: 3g; Iron: 1mg; Sodium: 280mg

Breakfast Pizza
PREP: 5 MINUTES • COOK TIME: 8 MINUTES • TOTAL: 13 MINUTES • SERVES: 1

Ingredients

2 large eggs
¼ cup unsweetened, unflavored almond milk
¼ teaspoon fine sea salt
⅛ teaspoon ground black pepper
¼ cup diced onions

¼ cup shredded Parmesan cheese
6 pepperoni slices
¼ teaspoon dried oregano leaves
¼ cup pizza sauce, warmed, for serving

Instructions

1   Preheat the Innsky air fryer to 350°F. Grease a 6 by 3-inch cake pan.
In a small bowl, use a fork to whisk together the eggs, almond milk, salt, and pepper. Add the onions and stir to mix. Pour the mixture into the greased pan. Top with the cheese (if using), pepperoni slices (if using), and oregano.

2   Place the pan in the air fryer and cook for 8 minutes, or until the eggs are cooked to your liking.
Loosen the eggs from the sides of the pan with a spatula and place them on a serving plate. Drizzle the pizza sauce on top. Best served fresh.

Per serving: Calories 357; Fat 25g; Protein 24g; Total carbs 9g; Fiber 2g

Easy Bacon
PREP: 2 MINUTES • COOK TIME: 6 MINUTES • TOTAL: 8 MINUTES • SERVES: 2

Ingredients
4 slices thin-cut bacon or beef bacon

Instructions
1 Spray the Innsky air fryer basket with avocado oil. Preheat the Innsky air fryer to 360°F.
2 Place the bacon in the Innsky air fryer basket in a single layer, spaced about ¼ inch apart Cook for 4 to 6 minutes (thicker bacon will take longer). Check the bacon after 4 minutes to make sure it is not overcooking. Best served fresh. Store extras in an airtight container in the fridge for up to 4 days. Reheat in a preheated 360°F air fryer for 2 minutes, or until heated through.

Per serving: Calories 140; Fat 12g; Protein 8g; Total carbs 0g; Fiber 0g

Blueberry Pancake Poppers
PREP: 5 MINUTES • COOK TIME: 8 MINUTES • TOTAL: 13 MINUTES • SERVES: 8

Ingredients
1 cup all-purpose flour
1 tablespoon sugar
1 teaspoon baking soda
½ teaspoon baking powder
1 cup milk

1 large egg
1 teaspoon vanilla extract
1 teaspoon olive oil
½ cup fresh blueberries

Instructions
1 In a medium mixing bowl, combine the flour, sugar, baking soda, and baking powder and mix well. Mix in the milk, egg, vanilla, and oil. Coat the inside of an air fryer muffin tin with cooking spray. Fill each muffin cup two-thirds full. (You may have to bake the poppers in more than one batch.) Drop a few blueberries into each muffin cup.
Set the muffin tin into the air fryer basket.
2 Set the temperature of your Innsky AF to 320°F. Set the timer and bake for 8 minutes. Insert a toothpick into the center of a pancake popper; if it comes out clean, they are done. If batter clings to the toothpick, cook the poppers for 2 minutes more and check again. When the poppers are cooked through, use silicone oven mitts to remove the muffin tin from the air fryer basket. Turn out the poppers onto a wire rack to cool.

Per Serving (1 popper): Calories: 103; Fat: 2g; Saturated fat: 1g; Carbohydrate: 18g; Fiber: 1g; Sugar: 4g; Protein: 4g; Iron: 1mg; Sodium: 181mg

Air Fryer Cinnamon Rolls
PREP: 5 MINUTES • COOK TIME: 12 MINUTES • TOTAL: 17 MINUTES • SERVES: 1

Ingredients

1 can of cinnamon rolls

Instructions

1   Spray the Innsky air fryer basket with olive oil.
    Separate the canned cinnamon rolls and place them in the air fryer basket.

2

    Set the temperature of your Innsky AF to 340°F. Set the timer and bake for 6 minutes.
    Using tongs, flip the cinnamon rolls. Reset the timer and bake for another 6 minutes.
    When the rolls are done cooking, use tongs to remove them from the air fryer. Transfer them to a platter and
    spread them with the icing that comes in the package.

Denver Omelet
PREP: 5 MINUTES • COOK TIME: 8 MINUTES • TOTAL: 13 MINUTES • SERVES: 1

Ingredients

2 large eggs

¼ cup unsweetened, unflavored almond milk

¼ teaspoon fine sea salt

⅛ teaspoon ground black pepper

¼ cup diced ham (omit for vegetarian)

¼ cup diced green and red bell peppers

2 tablespoons diced green onions, plus more for
  garnish

¼ cup shredded cheddar cheese

Quartered cherry tomatoes, for serving

Instructions

1   Preheat the Innsky air fryer to 350°F. Grease a 6 by 3-inch cake pan and set aside. In a small bowl, use a fork to
    whisk together the eggs, almond milk, salt, and pepper. Add the ham, bell peppers, and green onions. Pour the
    mixture into the greased pan. Add the cheese on top (if using).

2   Place the pan in the basket of the air fryer. Cook for 8 minutes, or until the eggs are cooked to your liking.
    Loosen the omelet from the sides of the pan with a spatula and place it on a serving plate. Garnish with green
    onions and serve with cherry tomatoes, if desired. Best served fresh.

Homemade Air Fried Banana Bread
PREP: 5 MINUTES • COOK TIME: 22 MINUTES • TOTAL: 13 MINUTES • SERVES: 3

Ingredients

3 ripe bananas, mashed

1 cup sugar

1 large egg

4 tablespoons (½ stick) unsalted butter, melted

1½ cups all-purpose flour

1 teaspoon baking soda

1 teaspoon salt

Instructions

1   Coat the insides of 3 mini loaf pans with cooking spray. In a large mixing bowl, mix together the bananas and
    sugar. In a separate large mixing bowl, combine the egg, butter, flour, baking soda, and salt and mix well. Add
    the banana mixture to the egg and flour mixture. Mix well. Divide the batter evenly among the prepared pans.
    Set the mini loaf pans into the air fryer basket.

2   Set the temperature of your Innsky AF to 310°F. Set the timer and bake for 22 minutes. Insert a toothpick into
    the center of each loaf; if it comes out clean, they are done. If the batter clings to the toothpick, cook the loaves
    for 2 minutes more and check again. When the loaves are cooked through, use silicone oven mitts to remove
    the pans from the air fryer basket. Turn out the loaves onto a wire rack to cool.

Air Fryer Homemade Blueberry Muffins
PREP: 5 MINUTES • COOK TIME: 14 MINUTES • TOTAL: 19 MINUTES • SERVES: 10

Ingredients

⅔ cup all-purpose flour
1 teaspoon baking powder
2 tablespoons sugar
1 egg

2 teaspoons vanilla extract
⅓ cup low-fat milk
3 tablespoons unsalted butter, melted
¾ cup fresh blueberries

Instructions

1. In a medium mixing bowl, combine the flour, baking powder, sugar, egg, vanilla, milk, and melted butter and mix well. Fold in the blueberries. Coat the inside of an air fryer muffin tin with cooking spray. Fill each muffin cup about two-thirds full. Set the muffin tin into the air fryer basket.

2. Set the temperature of your Innsky AF to 320°F. Set the timer and bake for 14 minutes.
Insert a toothpick into the center of a muffin; if it comes out clean, they are done. If batter clings to the toothpick, cook the muffins for 2 minutes more and check again. When the muffins are cooked through, use silicone oven mitts to remove the muffin tin from the air fryer basket. Turn out the muffins onto a wire rack to cool slightly before serving.

Per Serving: Calories: 92; Fat: 4g; Saturated fat: 2g; Carbohydrate: 12g; Fiber: 1g; Sugar: 4g; Protein: 2g; Iron: 1mg; Sodium: 35mg

Cheesy Baked Grits
PREP: 10 MINUTES • COOK TIME: 12 MINUTES • TOTAL: 22 MINUTES • SERVES: 6

Ingredients

¾ cup hot water
2 (1-ounce) packages instant grits (⅔ cup)
1 large egg, beaten
1 tablespoon butter, melted

2 cloves garlic, minced
½ to 1 teaspoon red pepper flakes
1 cup shredded cheddar cheese or jalapeño Jack cheese

Instructions

1. In a 6 × 3-inch round heatproof pan, combine the water, grits, egg, butter, garlic, and red pepper flakes. Stir until well combined. Stir in the shredded cheese.

2. Place the pan in the air fryer basket. Set the Innsky air fryer to 400°F for 12 minutes, or until the grits have cooked through and a knife inserted near the center comes out clean.
Let stand for 5 minutes before serving.

Easy Mexican Shakshuka
PREP: 5 MINUTES • COOK TIME: 6 MINUTES • TOTAL: 11 MINUTES • SERVES: 1

Ingredients

½ cup salsa
2 large eggs, room temperature
½ teaspoon fine sea salt
¼ teaspoon smoked paprika

⅛ teaspoon ground cumin
FOR GARNISH:
2 tablespoons cilantro leaves

Instructions

1. Preheat the Innsky air fryer to 400°F.
Place the salsa in a 6-inch pie pan or a casserole dish that will fit into your air fryer. Crack the eggs into the salsa and sprinkle them with the salt, paprika, and cumin.

2. Place the pan in the air fryer and cook for 6 minutes, or until the egg whites are set and the yolks are cooked to your liking. Remove from the air fryer and garnish with the cilantro before serving.

Per serving: Calories 258; Fat 17g; Protein 14g; Total carbs 11g; Fiber 4g

Breakfast Sammies

PREP: 15 MINUTES • COOK TIME: 20 MINUTES • TOTAL: 35 MINUTES • SERVES: 5

Ingredients

BISCUITS:

6 large egg whites

2 cups blanched almond flour, plus more if needed

1½ teaspoons baking powder

½ teaspoon fine sea salt

¼ cup (½ stick) very cold unsalted butter (or lard for dairy-free), cut into ¼-inch pieces

EGGS:

5 large eggs

½ teaspoon fine sea salt

¼ teaspoon ground black pepper

5 (1-ounce) slices cheddar cheese

10 thin slices ham, or 4 cooked Gyro Breakfast Patties

Instructions

1 Spray the Innsky air fryer basket with avocado oil. Preheat the Innsky air fryer to 350°F. Grease two 6-inch pie pans or two baking pans that will fit inside your air fryer.

Make the biscuits: In a medium-sized bowl, whip the egg whites with a hand mixer until very stiff. Set aside. In a separate medium-sized bowl, stir together the almond flour, baking powder, and salt until well combined. Cut in the butter. Gently fold the flour mixture into the egg whites with a rubber spatula. If the dough is too wet to form into mounds, add a few tablespoons of almond flour until the dough holds together well. Using a large spoon, divide the dough into 5 equal portions and drop them about 1 inch apart on one of the greased pie pans.

2 Place the pan in the air fryer and cook for 11 to 14 minutes, until the biscuits are golden brown. Remove from the air fryer and set aside to cool.

Make the eggs: Set the Innsky air fryer to 375°F. Crack the eggs into the remaining greased pie pan and sprinkle with the salt and pepper. Place the eggs in the air fryer to cook for 5 minutes, or until they are cooked to your liking. Open the air fryer and top each egg yolk with a slice of cheese (if using). Cook for another minute, or until the cheese is melted. Once the biscuits are cool, slice them in half lengthwise. Place 1 cooked egg topped with cheese and 2 slices of ham in each biscuit.

Store leftover biscuits, eggs, and ham in separate airtight containers in the fridge for up to 3 days. Reheat the biscuits and eggs on a baking sheet in a preheated 350°F air fryer for 5 minutes, or until warmed through.

Indian Masala Omelet

PREP: 10 MINUTES • COOK TIME: 12 MINUTES • TOTAL: 22 MINUTES • SERVES: 2

Ingredients

4 large eggs

½ cup diced onion

½ cup diced tomato

¼ cup chopped fresh cilantro

1 jalapeño, seeded and finely chopped

½ teaspoon ground turmeric

½ teaspoon kosher salt

½ teaspoon cayenne pepper

Olive oil for greasing the pan

Instructions

1 In a large bowl, beat the eggs. Stir in the onion, tomato, cilantro, jalapeño, turmeric, salt, and cayenne. Generously grease a 3-cup Bundt pan. (Be sure to grease the pan well—the proteins in eggs stick something fierce. And do not use a round baking pan. The hole in the center of a Bundt pan allows hot air to circulate through the middle of the omelet so that it will cook at the same rate as the outside.) Pour the egg mixture into the prepared pan.

2 Place the pan in the air fryer basket. Set the Innsky air fryer to 250°F for 12 minutes, or until the eggs are cooked through. Carefully unmold and cut the omelet into four pieces (2 pieces per serving).

Gyro Breakfast Patties with Tzatziki
PREP: 10 MINUTES • COOK TIME: 20 MINUTES • TOTAL: 13 MINUTES • SERVES: 16

Ingredients
PATTIES:
2 pounds ground lamb or beef
½ cup diced red onions
¼ cup sliced black olives
2 tablespoons tomato sauce
1 teaspoon dried oregano leaves
1 teaspoon Greek seasoning
2 cloves garlic, minced
1 teaspoon fine sea salt
TZATZIKI:
1 cup full-fat sour cream

1 small cucumber, chopped
½ teaspoon fine sea salt
½ teaspoon garlic powder, or 1 clove garlic, minced
¼ teaspoon dried dill weed, or 1 teaspoon finely chopped fresh dill
FOR GARNISH/SERVING:
½ cup crumbled feta cheese (about 2 ounces)
Diced red onions
Sliced black olives
Sliced cucumbers

Instructions
1 Preheat the Innsky air fryer to 350°F.
   Place the ground lamb, onions, olives, tomato sauce, oregano, Greek seasoning, garlic, and salt in a large bowl. Mix well to combine the ingredients. Using your hands, form the mixture into sixteen 3-inch patties.
2 Place about 5 of the patties in the air fryer and fry for 20 minutes, flipping halfway through. Remove the patties and place them on a serving platter. Repeat with the remaining patties.
   While the patties cook, make the tzatziki: Place all the ingredients in a small bowl and stir well. Cover and store in the fridge until ready to serve. Garnish with ground black pepper before serving.
   Serve the patties with a dollop of tzatziki, a sprinkle of crumbled feta cheese, diced red onions, sliced black olives, and sliced cucumbers. Store leftovers in an airtight container in the refrigerator for up to 5 days or in the freezer for up to a month. Reheat the patties in a preheated 390°F air fryer for a few minutes, until warmed through.

Per serving: Calories 396; Fat 31g; Protein 23g; Total carbs 4g; Fiber 0.4g

Queso Fundido
PREP: 10 MINUTES • COOK TIME: 25 MINUTES • TOTAL: 35 MINUTES • SERVES: 4

Ingredients
4 ounces fresh Mexican chorizo, casings removed
1 medium onion, chopped
3 cloves garlic, minced
1 cup chopped tomato
2 jalapeños, seeded and diced

2 teaspoons ground cumin
2 cups shredded Oaxaca or mozzarella cheese
½ cup half-and-half
Celery sticks or tortilla chips, for serving

Instructions
1 In a 6 × 3-inch round heatproof pan, combine the chorizo, onion, garlic, tomat jalapeños, and cumin. Stir to combine.
2 Place the pan in the air fryer basket. Set the Innsky air fryer to 400°F for 15 minutes, or until the sausage is cooked, stirring halfway through the cooking time to break up the sausage. Add the cheese and half-and-half; stir to combine. Set the Innsky air fryer to 325°F for 10 minutes, or until the cheese has melted. Serve with celery sticks or tortilla chips.

The Best Keto Quiche

PREP: 10 MINUTES • COOK TIME: 60 MINUTES • TOTAL: 70 MINUTES • SERVES: 1

Ingredients

CRUST:

1¼ cups blanched almond flour

1¼ cups grated Parmesan or Gouda cheese (about 3¾ ounces)

¼ teaspoon fine sea salt

1 large egg, beaten

FILLING:

½ cup chicken or beef broth (or vegetable broth for vegetarian)

1 cup shredded Swiss cheese (about 4 ounces)

4 ounces cream cheese (½ cup)

1 tablespoon unsalted butter, melted

4 large eggs, beaten

⅓ cup minced leeks or sliced green onions

¾ teaspoon fine sea salt

⅛ teaspoon cayenne pepper

Chopped green onions, for garnish

Instructions

1   Preheat the Innsky air fryer to 325°F. Grease a 6-inch pie pan. Spray two large pieces of parchment paper with avocado oil and set them on the countertop.

   Make the crust: In a medium-sized bowl, combine the flour, cheese, and salt and mix well. Add the egg and mix until the dough is well combined and stiff.

   Place the dough in the center of one of the greased pieces of parchment. Top with the other piece of parchment. Using a rolling pin, roll out the dough into a circle about 1/16 inch thick.

2   Press the pie crust into the prepared pie pan. Place it in the air fryer and bake for 12 minutes, or until it starts to lightly brown.

   While the crust bakes, make the filling: In a large bowl, combine the broth, Swiss cheese, cream cheese, and butter. Stir in the eggs, leeks, salt, and cayenne pepper. When the crust is ready, pour the mixture into the crust. Place the quiche in the air fryer and bake for 15 minutes. Turn the heat down to 300°F and bake for an additional 30 minutes, or until a knife inserted 1 inch from the edge comes out clean. You may have to cover the edges of the crust with foil to prevent burning.

   Allow the quiche to cool for 10 minutes before garnishing it with chopped green onions and cutting it into wedges.

   Store leftovers in an airtight container in the refrigerator for up to 4 days or in the freezer for up to a month. Reheat in a preheated 350°F air fryer for a few minutes, until warmed through.

Per serving: Calories 333; Fat 26g; Protein 20g; Total carbs 6g; Fiber 2g

Harissa Shakshuka
PREP: 15 MINUTES • COOK TIME: 15 MINUTES • TOTAL: 30 MINUTES • SERVES: 2

Ingredients
For the Harissa
½ cup olive oil
6 cloves garlic, minced
2 tablespoons smoked paprika
1 tablespoon ground coriander
1 tablespoon ground cumin
1 teaspoon ground caraway

1 teaspoon kosher salt
½ to 1 teaspoon cayenne pepper
For the Shakshuka
1 cup canned diced tomatoes with their liquid
4 large eggs
Chopped fresh parsley (optional)
Black pepper (optional)

Instructions
1 For the harissa: In a medium microwave-safe bowl, combine all the ingredients. Microwave on high for 1 minute, stirring halfway through the cooking time. (You can also heat this on the stovetop until the oil is hot and bubbling. Or, if you must use your air fryer for everything, cook in the air fryer at 350°F for 5 to 6 minutes, or until the paste is heated through.) For the shakshuka: In a 6 × 3-inch round heatproof pan, combine the tomatoes with 1 teaspoon of the harissa and stir until well combined. Taste and add more harissa if you want the sauce to be spicier.
   Carefully crack the eggs into the tomato mixture, taking care to not break the yolks.
2 Cover the pan with foil and place in the air fryer basket. Set the Innsky air fryer to 350°F for 15 minutes. Remove the foil. For a runny yolk, cook for an additional minutes; for a more set yolk, cook an additional 5 minutes. Garnish with fresh parsley and black pepper, if desired.

Green Eggs and Ham
PREP: 5 MINUTES • COOK TIME: 10 MINUTES • TOTAL: 15 MINUTES • SERVES: 2

Ingredients
1 large Hass avocado, halved and pitted
2 thin slices ham
2 large eggs
2 tablespoons chopped green onions, plus more for
  garnish

½ teaspoon fine sea salt
¼ teaspoon ground black pepper
¼ cup shredded cheddar cheese (omit for dairy-free)

Instructions
1 Preheat the Innsky air fryer to 400°F.
   Place a slice of ham into the cavity of each avocado half. Crack an egg on top of the ham, then sprinkle on the green onions, salt, and pepper.
2 Place the avocado halves in the air fryer cut side up and cook for 10 minutes, or until the egg is cooked to your desired doneness. Top with the cheese (if using) and cook for 30 seconds more, or until the cheese is melted. Garnish with chopped green onions.
   Best served fresh. Store extras in an airtight container in the fridge for up to 4 days. Reheat in a preheated 350°F air fryer for a few minutes, until warmed through.

Per serving: Calories 307; Fat 24g; Protein 14g; Total carbs 10g; Fiber 7g

Toad In The Hole

PREP: 10 MINUTES • COOK TIME: 35 MINUTES • TOTAL: 45 MINUTES • SERVES: 4

Ingredients

½ cup all-purpose flour

½ teaspoon kosher salt

½ teaspoon black pepper

4 large eggs

1 cup whole milk

2 tablespoons Dijon mustard

2 tablespoons vegetable oil

4 uncooked pork sausages (about 4 ounces each)

Instructions

1   In a medium bowl, whisk together the flour, salt, and pepper. Make a well in the middle. In a separate medium bowl, combine the eggs, milk, and mustard and whisk until well blended. Slowly whisk the egg mixture into the flour. (You want a batter that's about as thick as pancake batter. If it's too thick, add a little water or additional milk.) Cover the batter and let it rest while you cook the sausages.

Pour the oil into a 6 × 3-inch round heatproof pan. Cut the sausages in half and place in the pan.

2   Place the pan in the air fryer basket. Set the Innsky air fryer to 400°F for 15 minutes.

Carefully pour the batter on top of the sausages. Set the Innsky air fryer to 350°F for 20 minutes, or until the batter has puffed up and is browned on top. Cut into 4 wedges and serve hot.

Everything Bagels

PREP: 15 MINUTES • COOK TIME: 14 MINUTES • TOTAL: 29 MINUTES • SERVES: 6

Ingredients

1¾ cups shredded mozzarella cheese or goat cheese mozzarella

2 tablespoons unsalted butter or coconut oil

1 large egg, beaten

1 tablespoon apple cider vinegar

1 cup blanched almond flour

1 tablespoon baking powder

⅛ teaspoon fine sea salt

1½ teaspoons everything bagel seasoning

Instructions

1   Make the dough: Put the mozzarella and butter in a large microwave-safe bowl and microwave for 1 to 2 minutes, until the cheese is entirely melted. Stir well. Add the egg and vinegar. Using a hand mixer on medium, combine well. Add the almond flour, baking powder, and salt and, using the mixer, combine well. Lay a piece of parchment paper on the countertop and place the dough on it. Knead it for about 3 minutes. The dough should be a little sticky but pliable.

Preheat the Innsky air fryer to 350°F. Spray a baking sheet or pie pan that will fit into your air fryer with avocado oil. Divide the dough into 6 equal portions. Roll 1 portion into a log that is 6 inches long and about ½ inch thick. Form the log into a circle and seal the edges together, making a bagel shape. Repeat with the remaining portions of dough, making 6 bagels.

Place the bagels on the greased baking sheet. Spray the bagels with avocado oil and top with everything bagel seasoning, pressing the seasoning into the dough with your hands.

2   Place the bagels in the air fryer and cook for 14 minutes, or until cooked through and golden brown, flipping after 6 minutes. Remove the bagels from the air fryer and allow them to cool slightly before slicing them in half and serving. Store leftovers in an airtight container in the fridge for up to 4 days or in the freezer for up to a month. Bagel chips are a great snack, and they're easy to make with these bagels. Slice a bagel into ¼-inch-thick rounds and spread each round with 1 teaspoon unsalted butter. Place the rounds in the air fryer in a single layer, leaving space between them, and cook for 5 to 6 minutes, until crispy. They're best served fresh, but they'll keep in an airtight container in the fridge for up to 4 days or in the freezer for up to a month.

Per serving: Calories 224; Fat 19g; Protein 12g; Total carbs 4g; Fiber 2g

Breakfast Cobbler

PREP: 20 MINUTES • COOK TIME: 30 MINUTES • TOTAL: 50 MINUTES • SERVES: 4

Ingredients

FILLING:

10 ounces bulk pork sausage, crumbled

¼ cup minced onions

2 cloves garlic, minced

½ teaspoon fine sea salt

½ teaspoon ground black pepper

1 (8-ounce) package cream cheese softened

¾ cup beef or chicken broth

BISCUITS:

3 large egg whites

¾ cup blanched almond flour

1 teaspoon baking powder

¼ teaspoon fine sea salt

2½ tablespoons very cold unsalted butter, cut into ¼-inch pieces

Fresh thyme leaves, for garnish

Instructions

1 Preheat the Innsky air fryer to 400°F.

Place the sausage, onions, and garlic in a 7-inch pie pan. Using your hands, break up the sausage into small pieces and spread it evenly throughout the pie pan. Season with the salt and pepper.

2 Place the pan in the air fryer and cook for 5 minutes.

While the sausage cooks, place the cream cheese and broth in a food processor or blender and puree until smooth.

Remove the pork from the air fryer and use a fork or metal spatula to crumble it more. Pour the cream cheese mixture into the sausage and stir to combine. Set aside.

Make the biscuits: Place the egg whites in a medium-sized mixing bowl or the bowl of a stand mixer and whip with a hand mixer or stand mixer until stiff peaks form.

In a separate medium-sized bowl, whisk together the almond flour, baking powder, and salt, then cut in the butter. When you are done, the mixture should still have chunks of butter. Gently fold the flour mixture into the egg whites with a rubber spatula.

Use a large spoon or ice cream scoop to scoop the dough into 4 equal-sized biscuits, making sure the butter is evenly distributed. Place the biscuits on top of the sausage and cook in the air fryer for 5 minutes, then turn the heat down to 325°F and cook for another 17 to 20 minutes, until the biscuits are golden brown. Serve garnished with fresh thyme leaves.

Store leftovers in an airtight container in the refrigerator for up to 3 days. Reheat in a preheated 350°F air fryer for 5 minutes, or until warmed through.

Per serving: Calories 623; Fat 55g; Protein 23g; Total carbs 8g; Fiber 3g

Keto Danish

PREP: 15 MINUTES • COOK TIME: 20 MINUTES • TOTAL: 13 MINUTES • SERVES: 6

Ingredients

3 large eggs
¼ teaspoon cream of tartar
¼ cup vanilla-flavored egg white protein powder
¼ cup Swerve confectioners'-style sweetener
3 tablespoons full-fat sour cream
1 teaspoon vanilla extract
FILLING:
4 ounces cream cheese (½ cup)
2 large egg yolks (from above)

¼ cup Swerve confectioners'-style sweetener
1 teaspoon vanilla extract
¼ teaspoon ground cinnamon
DRIZZLE:
1 ounce cream cheese (2 tablespoons)
1 tablespoon Swerve confectioners'-style sweetener
1 tablespoon unsweetened, unflavored almond milk
(or heavy cream for nut-free)

Instructions

1 Preheat the Innsky air fryer to 300°F. Spray a casserole dish that will fit in your air fryer with avocado oil.
   Make the pastry: Separate the eggs, putting all the whites in a large bowl, one yolk in a medium-sized bowl, and two yolks in a small bowl. Beat all the egg yolks and set aside.
   Add the cream of tartar to the egg whites. Whip the whites with a hand mixer until very stiff, then turn the hand mixer's setting to low and slowly add the protein powder while mixing. Mix until only just combined; if you mix too long, the whites will fall. Set aside.
   To the egg yolk in the medium-sized bowl, add the sweetener, sour cream, and vanilla extract. Mix well. Slowly pour the yolk mixture into the egg whites and gently combine. Dollop 6 equal-sized mounds of batter into the casserole dish. Use the back of a large spoon to make an indentation on the top of each mound. Set aside.
   Make the filling: Place the cream cheese in a small bowl and stir to break it up. Add the 2 remaining egg yolks, the sweetener, vanilla extract, and cinnamon and stir until well combined. Divide the filling among the mounds of batter, pouring it into the indentations on the tops.

2 Place the Danish in the air fryer and bake for about 20 minutes, or until golden brown.
   While the Danish bake, make the drizzle: In a small bowl, stir the cream cheese to break it up. Stir in the sweetener and almond milk. Place the mixture in a piping bag or a small resealable plastic bag with one corner snipped off. After the Danish have cooled, pipe the drizzle over the Danish.
   Store leftovers in airtight container in the fridge for up to 4 days.

Per serving: Calories 160; Fat 12g; Protein 8g; Total carbs 2g; Fiber 0.3g

French Toast Pavlova

PREP:15 MINUTES PLUS 20 MINUTES TO REST•COOK TIME: 60 MINUTES • TOTAL: 1 HOUR 35 MINUTES•SERVES: 1

Ingredients

3 large egg whites

¼ teaspoon cream of tartar

¾ cup Swerve confectioners'-style sweetener or equivalent amount of powdered sweetener

1 teaspoon ground cinnamon

1 teaspoon maple extract

TOPPINGS:

½ cup heavy cream

3 tablespoons Swerve confectioners'-style sweetener or equivalent amount of powdered sweetener, plus more for garnish

Fresh strawberries (optional)

Instructions

1 Preheat the Innsky air fryer to 275°F. Thoroughly grease a 7-inch pie pan with butter or coconut oil. Place a large bowl in the refrigerator to chill.

In a small bowl, combine the egg whites and cream of tartar. Using a hand mixer, beat until soft peaks form. Turn the mixer to low and slowly sprinkle in the sweetener while mixing until completely incorporated. Add the cinnamon and maple extract and beat on medium-high until the peaks become stiff.

Spoon the mixture into the greased pie pan, then smooth it across the bottom, up the sides, and onto the rim of the pie pan to form a shell.

2 Cook in the air fryer for 1 hour, then turn off the air fryer and let the shell stand in the air fryer for another 20 minutes. Once the shell has set, transfer it to the refrigerator to chill for 20 minutes or the freezer to chill for 10 minutes.

While the shell sets and chills, make the topping: Remove the large bowl from the refrigerator and place the heavy cream in it. Whip with a hand mixer on high until soft peaks form. Add the sweetener and beat until medium peaks form. Taste and adjust the sweetness to your liking.

Place the chilled shell on a serving platter and spoon on the cream topping. Top with the strawberries, if desired, and garnish with powdered sweetener. Slice and serve.

If you won't be eating the pavlova right away, store the shell and topping in separate airtight containers in the refrigerator for up to 3 days.

Per serving: Calories 115; Fat 11g; Protein 3g; Total carbs 2g; Fiber 0.3g

Bacon-Wrapped Pickle Spears
PREP: 10 MINUTES • COOK TIME: 8 MINUTES • TOTAL: 18 MINUTES •SERVES: 4

Ingredients

8 to 12 slices bacon

¼ cup (2 ounces) cream cheese, softened

¼ cup shredded mozzarella cheese

8 dill pickle spears

½ cup ranch dressing

Instructions

1. Lay the bacon slices on a flat surface. In a medium bowl, combine the cream cheese and mozzarella. Stir until well blended. Spread the cheese mixture over the bacon slices.
Place a pickle spear on a bacon slice and roll the bacon around the pickle in a spiral, ensuring the pickle is fully covered. (You may need to use more than one slice of bacon per pickle to fully cover the spear.) Tuck in the ends to ensure the bacon stays put. Repeat to wrap all the pickles.

2. Place the wrapped pickles in the Innsky air fryer basket in a single layer. Set the Innsky air fryer to 400°F for 8 minutes, or until the bacon is cooked through and crisp on the edges.
Serve the pickle spears with ranch dressing on the side.

Buffalo Cauliflower
PREP: 5 MINUTES • COOK TIME: 11 MINUTES • TOTAL: 16 MINUTES
SERVES: 4

Ingredients

¼ cup hot sauce

¼ cup powdered Parmesan cheese

2 tablespoons unsalted butter, melted

1 small head cauliflower, cut into 1-inch bites

Blue Cheese Dressing, for serving

Blue Cheese Dressing
PREP: 20 MINUTES • COOK TIME: 30 MINUTES • TOTAL: 50 MINUTES
SERVES: 2 CUPS

Ingredients

8 ounces crumbled blue cheese, plus more if desired
  for a chunky texture

¼ cup beef bone broth

¼ cup full-fat sour cream

¼ cup red wine vinegar or coconut vinegar

1½ tablespoons Swerve confectioners'-style
  sweetener or equivalent amount of liquid or
  powdered sweetener

1 tablespoon MCT oil

1clove garlic, peeled

Instructions

1. Place all the ingredients in a food processor and blend until smooth. Transfer to a jar. Stir in extra chunks of blue cheese if desired. Store in the refrigerator for up to 5 days. Preheat the Innsky air fryer to 400°F. Spray a baking dish that will fit into your air fryer with avocado oil.
Place the hot sauce, Parmesan, and butter in a large bowl and stir to combine. Add the cauliflower and toss to coat well.

2. Place the coated cauliflower in the baking dish. Cook in the air fryer for 11 minutes, stirring halfway through. Serve with blue cheese dressing.
Store leftovers in an airtight container in the fridge for up to 4 days. Reheat in a preheated 400°F oven for 3 minutes, until warmed through and crispy.

Per serving: Calories 185; Fat 15g; Protein 9g; Total carbs 4g; Fiber 2g

Ranch Roasted Chickpeas
PREP: 4 MINUTES • COOK TIME: 10 MINUTES • TOTAL: 14 MINUTES • SERVES: 4

Ingredients
1 (15-ounce) can chickpeas, drained and rinsed
1 tablespoon olive oil
3 tablespoons ranch seasoning mix

1 teaspoon salt
2 tablespoons freshly squeezed lemon juice

Instructions
1. Spray the Innsky air fryer basket with olive oil.
   Using paper towels, pat the chickpeas dry. In a medium mixing bowl, mix together the chickpeas, oil, seasoning mix, salt, and lemon juice. Put the chickpeas in the Innsky air fryer basket and spread them out in a single layer. (You may need to cook the chickpeas in more than one batch.)
2. Set the temperature of your Innsky AF to 350°F. Set the timer and roast for 4 minutes. Remove the drawer and shake vigorously to redistribute the chickpeas so they cook evenly. Reset the timer and roast for 6 minutes more. When the time is up, release the Innsky air fryer basket from the drawer and pour the chickpeas into a bowl. Season with additional salt, if desired. Enjoy!

Per Serving: Calories: 144; Fat: 5g; Saturated fat: 1g; Carbohydrate: 19g; Fiber: 5g; Sugar: 3g; Protein: 6g; Iron: 2mg; Sodium: 891mg

Caramelized Onion Dip
PREP:5 MINUTES PLUS 2 HOURS TO CHILL•COOK TIME: 30 MINUTES •TOTAL: 2 HOUR 35 MINUTES • SERVES: 8-10

Ingredients
1 tablespoon butter
1 medium yellow onion, halved and thinly sliced
¼ teaspoon kosher salt, plus additional for seasoning
4 ounces cream cheese, softened
½ cup sour cream

¼ teaspoon onion powder
1 tablespoon chopped fresh chives
Black pepper
Thick-cut potato chips or vegetable chips

Instructions
1. Place the butter in a 6 × 3-inch round heatproof pan. Place the pan in the air fryer basket.
2. Set the Innsky air fryer to 200°F for 1 minute, or until the butter is melted. Add the onions and salt to the pan. Set the Innsky air fryer to 200°F for 15 minutes, or until onions are softened. Set the Innsky air fryer to 375°F for 15 minutes, until onions are a deep golden brown, stirring two or three times during the cooking time. Let cool completely. In a medium bowl, stir together the cooked onions, cream cheese, sour cream, onion powder, and chives. Season with salt and pepper. Cover and refrigerate for 2 hours to allow the flavors to blend. Serve the dip with potato chips or vegetable chips.

Homemade Air Fryer Roasted Mixed Nuts
PREP: 5 MINUTES • COOK TIME: 20 MINUTES • TOTAL: 25 MINUTES • SERVES: 6

Ingredients
2 cups mixed nuts (walnuts, pecans, and/or almonds)
2 tablespoons egg white
1 teaspoon ground cinnamon

2 tablespoons sugar
1 teaspoon paprika

Instructions
1. Preheat the Innsky air fryer to 300°F and spray the Innsky air fryer basket with olive oil. In a small mixing bowl, mix together the nuts, egg white, cinnamon, sugar, and paprika, until the nuts are thoroughly coated.
2. Place the nuts in the greased air fryer basket; set the timer and roast for 10 minutes. After 10 minutes, remove the drawer and shake the basket to redistribute the nuts so they roast evenly. Reset the timer and roast for 10 minutes more. Release the basket from the drawer, pour the nuts into a bowl, and serve.

Per Serving: Calories: 232; Fat: 21g; Saturated fat: 2g; Carbohydrate: 10g; Fiber: 3g; Sugar: 5g; Protein: 6g; Iron: 1mg; Sodium: 6mg

Ranch Kale Chips
PREP: 5 MINUTES • COOK TIME: 10 MINUTES • TOTAL: 15 MINUTES • SERVES: 8 CUPS

Ingredients
½ teaspoon dried chives
½ teaspoon dried dill weed
½ teaspoon dried parsley
¼ teaspoon garlic powder

¼ teaspoon onion powder
⅛ teaspoon fine sea salt
⅛ teaspoon ground black pepper
2 large bunches kale

Instructions
1. Spray the Innsky air fryer basket with avocado oil. Preheat the Innsky air fryer to 360°F.
   Place the seasonings, salt, and pepper in a small bowl and mix well.
   Wash the kale and pat completely dry. Use a sharp knife to carve out the thick inner stems, then spray the leaves with avocado oil and sprinkle them with the seasoning mix.
2. Place the kale leaves in the air fryer in a single layer and cook for 10 minutes, shaking and rotating the chips halfway through. Transfer the baked chips to a baking sheet to cool completely and crisp up. Repeat with the remaining kale. Sprinkle the cooled chips with salt before serving, if desired.
   Kale chips can be stored in an airtight container at room temperature for up to 1 week, but they are best eaten within 3 days.

Per serving: Calories 11; Fat 0.2g; Protein 1g; Total carbs 2g; Fiber 0.4g

Cheese Drops

PREP: 15 MINUTES • COOK TIME: 10 MINUTES • TOTAL: 25 MINUTES • SERVES: 4

Ingredients

¾ cup all-purpose flour
½ teaspoon kosher salt
¼ teaspoon cayenne pepper
¼ teaspoon smoked paprika
¼ teaspoon black pepper

Dash garlic powder (optional)
¼ cup butter, softened
1 cup shredded sharp cheddar cheese, at room temperature
Olive oil spray

Instructions

1. In a small bowl, combine the flour, salt, cayenne, paprika, pepper, and garlic powder, if using. Using a food processor, cream the butter and cheese until smooth. Gently add the seasoned flour and process until the dough is well combined, smooth, and no longer sticky. (Or make the dough in a stand mixer fitted with the paddle attachment: Cream the butter and cheese on medium speed until smooth, then add the seasoned flour and beat at low speed until smooth.)
   Divide the dough into 32 equal-size pieces. On a lightly floured surface, roll each piece into a small ball.
2. Spray the Innsky air fryer basket with oil spray. Arrange 16 cheese drops in the basket. Set the Innsky air fryer to 325°F for 10 minutes, or until drops are just starting to brown. Transfer to a wire rack. Repeat with remaining dough, checking for doneness at 8 minutes. Cool the cheese drops completely on the wire rack. Store in an airtight container until ready to serve, or up to 1 or 2 days.

Onion Pakoras

PREP: 5 MINUTES PLUS 30 MINUTES TO STAND • COOK TIME: 10 MINUTES • TOTAL: 45 MINUTES • SERVES: 4

Ingredients

2 medium yellow or white onions, sliced
½ cup chopped fresh cilantro
2 tablespoons vegetable oil
1 tablespoon chickpea flour
1 tablespoon rice flour

1 teaspoon ground turmeric
1 teaspoon cumin seeds
1 teaspoon kosher salt
½ teaspoon cayenne pepper
Vegetable oil spray

Instructions

1. In a large bowl, combine the onions, cilantro, oil, chickpea flour, rice flour, turmeric, cumin seeds, salt, and cayenne. Stir to combine. Cover and let stand for 30 minutes or up to overnight. (This allows the onions to release moisture, creating a batter.) Mix well before using. Spray the Innsky air fryer basket generously with vegetable oil spray. Drop half of the batter in 6 heaping tablespoons into the basket.
2. Set the Innsky air fryer to 350°F for 8 minutes. Carefully turn the pakoras over and spray with oil spray. Set the Innsky air fryer for 2 minutes, or until the batter is cooked through and crisp. Repeat with remaining batter to make 6 more pakoras, checking at 6 minutes for doneness. Serve hot.

Lebanese Muhammara

PREP: 15 MINUTES • COOK TIME: 15 MINUTES • TOTAL: 30 MINUTES • SERVES: 6

Ingredients

2 large red bell peppers
¼ cup plus 2 tablespoons olive oil
1 cup walnut halves
1 tablespoon agave nectar or honey
1 teaspoon fresh lemon juice
1 teaspoon ground cumin

1 teaspoon kosher salt
1 teaspoon red pepper flakes
Raw vegetables (such as cucumber, carrots, zucchini slices, or cauliflower) or toasted pita chips, for serving

Instructions

1. Drizzle the peppers with 2 tablespoons of the olive oil and place in the air fryer basket. Set the Innsky air fryer to 400°F for 10 minutes.

2. Add the walnuts to the basket, arranging them around the peppers. Set the Innsky air fryer to 400°F for 5 minutes. Remove the peppers, seal in a resealable plastic bag, and let rest for 5 to 10 minutes. Transfer the walnuts to a plate and set aside to cool.

   Place the softened peppers, walnuts, agave, lemon juice, cumin, salt, and ½ teaspoon of the pepper flakes in a food processor and puree until smooth. Transfer the dip to a serving bowl and make an indentation in the middle. Pour the remaining ¼ cup olive oil into the indentation. Garnish the dip with the remaining ½ teaspoon pepper flakes.

   Serve with vegetables or toasted pita chips.

Easy Tomato And Basil Bruschetta

PREP: 5 MINUTES • COOK TIME: 3 MINUTES • TOTAL: 8 MINUTES • SERVES: 6

Ingredients

4 tomatoes, diced
⅓ cup fresh basil, shredded
¼ cup shredded Parmesan cheese
1 tablespoon minced garlic
1 tablespoon balsamic vinegar

1 teaspoon olive oil
1 teaspoon salt
1 teaspoon freshly ground black pepper
1 loaf French bread

Instructions

In a medium mixing bowl, combine the tomatoes and basil.

Mix in the Parmesan cheese, garlic, vinegar, olive oil, salt, and pepper. Let the tomato mixture sit and marinate, while you prepare the bread. Spray the Innsky air fryer basket with olive oil. Cut the bread into 1-inch-thick slices. Place the slices in the greased Innsky air fryer basket in a single layer. Spray the top of the bread with olive oil. Set the temperature of your Innsky AF to 250°F. Set the timer and toast for 3 minutes.

Using tongs, remove the bread slices from the air fryer and place a spoonful of the bruschetta topping on each piece.

Per Serving: Calories: 258; Fat: 3g; Saturated fat: 1g; Carbohydrate: 47g; Fiber: 3g; Sugar: 4g; Protein: 11g; Iron: 3mg; Sodium: 826mg

Crispy Nacho Avocado Fries
PREP: 10 MINUTES • COOK TIME: 15 MINUTES • TOTAL: 25 MINUTES • SERVES: 6

Ingredients

3 firm, barely ripe avocados, halved, peeled, and pitted

2 cups pork dust (or powdered Parmesan cheese for vegetarian)

2 teaspoons fine sea salt

2 teaspoons ground black pepper

2 teaspoons ground cumin

1 teaspoon chili powder

1 teaspoon paprika

½ teaspoon garlic powder

½ teaspoon onion powder

2 large eggs

Salsa, for serving (optional)

Fresh chopped cilantro leaves, for garnish (optional)

Instructions

1. Spray the Innsky air fryer basket with avocado oil. Preheat the Innsky air fryer to 400°F.

   Slice the avocados into thick-cut french fry shapes.

   In a bowl, mix together the pork dust, salt, pepper, and seasonings.

   In a separate shallow bowl, beat the eggs.

   Dip the avocado fries into the beaten eggs and shake off any excess, then dip them into the pork dust mixture. Use your hands to press the breading into each fry.

   Spray the fries with avocado oil and place them in the Innsky air fryer basket in a single layer, leaving space between them. If there are too many fries to fit in a single layer, work in batches.

2. Cook in the air fryer for 13 to 15 minutes, until golden brown, flipping after 5 minutes.

   Serve with salsa, if desired, and garnish with fresh chopped cilantro, if desired. Best served fresh.

   Store leftovers in an airtight container in the fridge for up to 5 days. Reheat in a preheated 400°F air fryer for 3 minutes, or until heated through.

Per serving: Calories 282; Fat 22g; Protein 15g; Total carbs 9g; Fiber 7g

Seasoned Sausage Rolls
PREP: 5 MINUTES • COOK TIME: 5 MINUTES • TOTAL: 10 MINUTES • SERVES: 6

Ingredients

FOR THE SEASONING

2 tablespoons sesame seeds

1½ teaspoons poppy seeds

1½ teaspoons dried minced onion

1 teaspoon salt

1 teaspoon dried minced garlic

FOR THE SAUSAGES

1 (8-ounce) package crescent roll dough

1 (12-ounce) package mini smoked sausages (cocktail franks)

Instructions

1. TO MAKE THE SEASONING: In a small bowl, combine the sesame seeds, poppy seeds, onion, salt, and garlic and set aside.

   TO MAKE THE SAUSAGES. Spray the Innsky air fryer basket with olive oil. Remove the crescent dough from the package and lay it out on a cutting board. Separate the dough at the perforations. Using a pizza cutter or sharp knife, cut each triangle of dough into fourths. Drain the sausages and pat them dry with a paper towel. Roll each sausage in a piece of dough. Sprinkle seasoning on top of each roll. Place the seasoned sausage rolls into the greased Innsky air fryer basket in a single layer.

2. Set the temperature of your Innsky AF to 330°F. Set the timer for 5 minutes. Using tongs, remove the sausages from the air fryer and place them on a platter. Repeat steps 6 through 8 with the second batch.

   Per Serving: Calories: 344; Fat: 26g; Saturated fat: 8g; Carbohydrate: 17g; Fiber: 1g; Sugar: 3g; Protein: 10g; Iron: 2mg; Sodium: 1145mg

Homemade Air Fryer Pita Chips

PREP: 5 MINUTES • COOK TIME: 6 MINUTES • TOTAL: 11 MINUTES • SERVES: 4

Ingredients

2 pieces whole wheat pita bread
3 tablespoons olive oil
1 teaspoon freshly squeezed lemon juice

1 teaspoon salt
1 teaspoon dried basil
1 teaspoon garlic powder

Instructions

1. Spray the Innsky air fryer basket with olive oil. Using a pair of kitchen shears or a pizza cutter, cut the pita bread into small wedges. Place the wedges in a small mixing bowl and add the olive oil, lemon juice, salt, dried basil, and garlic powder. Mix well, coating each wedge. Place the seasoned pita wedges in the greased Innsky air fryer basket in a single layer, being careful not to overcrowd them.
   Set the temperature of your Innsky AF to 350°F. Set the timer and bake for 6 minutes. Every 2 minutes or so, remove the drawer and shake the pita chips so they redistribute in the basket for even cooking. Serve with your choice of dip or alone as a tasty snack.

Bacon-Wrapped Pickle Poppers

PREP: 10 MINUTES • COOK TIME: 10 MINUTES • TOTAL: 20 MINUTES • SERVES: 24 POPPERS

Ingredients

12 medium dill pickles
1 (8-ounce) package cream cheese, softened
1 cup shredded sharp cheddar cheese

12 slices bacon or beef bacon, sliced in half lengthwise
Ranch Dressing or Blue Cheese Dressing, for serving (optional)

Instructions

1. Spray the Innsky air fryer basket with avocado oil. Preheat the Innsky air fryer to 400°F. Slice the dill pickles in half lengthwise and use a spoon to scoop out the centers. Place the cream cheese and cheddar cheese in a small bowl and stir until well combined. Divide the cream cheese mixture among the pickles, spooning equal amounts into the scooped-out centers. Wrap each filled pickle with a slice of bacon and secure the bacon with toothpicks.
2. Place the bacon-wrapped pickles in the Innsky air fryer basket with the bacon seam side down and cook for 8 to 10 minutes, until the bacon is crispy, flipping halfway through. Serve warm with ranch or blue cheese dressing, if desired.
   Best served fresh. Store leftovers in an airtight container in the fridge for up to 5 days. Reheat in a preheated 400°F air fryer for 3 minutes, or until heated through.

Per serving: Calories 87; Fat 8g; Protein 4g; Total carbs 1g; Fiber 1g

Doro Wat Wings

PREP: 5 MINUTES • COOK TIME: 32 MINUTES • TOTAL: 37 MINUTES • SERVES: 1 DOZEN WINGS

Ingredients

1 dozen chicken wings or drummies
1 tablespoon coconut oil or bacon fat, melted
2 teaspoons berbere spice
1 teaspoon fine sea salt
FOR SERVING

(OMIT FOR EGG-FREE):
2 hard-boiled eggs
½ teaspoon fine sea salt
¼ teaspoon berbere spice
¼ teaspoon dried chives

Instructions

1. Spray the Innsky air fryer basket with avocado oil. Preheat the Innsky air fryer to 380°F.
   Place the chicken wings in a large bowl. Pour the oil over them and turn to coat completely. Sprinkle the berbere and salt on all sides of the chicken.

2. Place the chicken wings in the air fryer and cook for 25 minutes, flipping after 15 minutes.
   After 25 minutes, increase the temperature to 400°F and cook for 6 to 7 minutes more, until the skin is browned and crisp.
   While the chicken cooks, prepare the hard-boiled eggs (if using): Peel the eggs, slice them in half, and season them with the salt, berbere, and dried chives. Serve the chicken and eggs together.

Per serving: Calories 317; Fat 24g; Protein 24g; Total carbs 0.1g; Fiber 0g

Bourbon Chicken Wings

PREP: 10 MINUTES • COOK TIME: 32 MINUTES • TOTAL: 42 MINUTES • SERVES: 8

Ingredients

2 pounds chicken wings or drummies
½ teaspoon fine sea salt
SAUCE:
½ cup chicken broth
⅓ cup Swerve confectioners'-style sweetener or equivalent amount of liquid or powdered sweetener
¼ cup tomato sauce
¼ cup wheat-free tamari

1 tablespoon apple cider vinegar
¾ teaspoon red pepper flakes
¼ teaspoon grated fresh ginger
1 clove garlic, smashed to a paste
FOR GARNISH (OPTIONAL):
Chopped green onions
Sesame seeds

Instructions

1. Spray the Innsky air fryer basket with avocado oil. Preheat the Innsky air fryer to 380°F.
   Season the chicken wings on all sides with the salt and place them in the air fryer.

2. Cook for 25 minutes, flipping after 15 minutes. After 25 minutes, increase the temperature to 400°F and cook for 6 to 7 minutes more, until the skin is browned and crisp.

Masala Peanuts
PREP: 15 MINUTES • COOK TIME: 15 MINUTES • TOTAL: 50 MINUTES • SERVES: 4

Ingredients
5 tablespoons chickpea flour
½ teaspoon cumin seeds
¼ teaspoon ground turmeric
¼ teaspoon kosher salt, plus more for seasoning if desired
¼ to ½ teaspoon cayenne pepper
2 tablespoons vegetable oil

3 tablespoons water
1 cup red Spanish peanuts or unsalted roasted peanuts
Vegetable oil spray
Prepared chaat masala or amchoor (dried mango powder), optional

Instructions
1. In a medium bowl, combine the chickpea flour, cumin seeds, turmeric, salt, and cayenne. Add the oil and stir to combine. Add the water and stir to make a thick, pancake-like batter. Add the peanuts and stir until well blended. Place a circle of parchment paper in the bottom of the Innsky air fryer basket. Pour the peanut mixture onto the parchment paper.
2. Set the Innsky air fryer to 325°F for 10 minutes.
   Open the air fryer and break up the peanuts and batter. Remove the parchment paper and let the peanuts sit directly on the bottom of the air fryer basket. Spray the peanuts generously with the vegetable oil spray. (Don't skip this step or the batter will taste raw.) Set the Innsky air fryer to 400°F for 5 minutes, or until the outsides of the peanuts are crisp.
   Transfer the peanuts to a rimmed baking sheet and shake well. Sprinkle with chaat masala or amchoor, if using. (You can also sprinkle with a little kosher salt.) Let peanuts cool for 10 minutes before serving. (They will continue to crisp as they cool.) Store in an airtight container.

Easy Mozzarella Sticks
PREP: 10 MINUTES • COOK TIME: 8 MINUTES • TOTAL: 18 MINUTES • SERVES: 6

Ingredients
1 (12-count) package mozzarella sticks
1 (8-ounce) package crescent roll dough
3 tablespoons unsalted butter, melted

¼ cup panko bread crumbs
Marinara sauce, for dipping (optional)

Instructions
1. Spray the Innsky air fryer basket with olive oil.
   Cut each cheese stick into thirds. Unroll the crescent roll dough. Using a pizza cutter or sharp knife, cut the dough into 36 even pieces.
   Wrap each small cheese stick in a piece of dough. Make sure that the dough is wrapped tightly around the cheese. Pinch the dough together at both ends, and pinch along the seam to ensure that the dough is completely sealed. Using tongs, dip the wrapped cheese sticks in the melted butter, then dip the cheese sticks in the panko bread crumbs.
   Place the cheese sticks in the greased Innsky air fryer basket in a single layer. (You may have to cook the cheese sticks in more than one batch.)
2. Set the temperature of your Innsky AF to 370°F. Set the timer and bake for 5 minutes. After 5 minutes, the tops should be golden brown. Using tongs, flip the cheese sticks and bake for another 3 minutes, or until golden brown on all sides. Repeat until you use all of the dough. Plate, serve with the marinara sauce, and enjoy.

Per Serving: Calories: 348; Fat: 23g; Saturated fat: 13g; Carbohydrate: 21g; Fiber: 1g; Sugar: 3g; Protein: 17g; Iron: 1mg; Sodium: 811mg

Pepperoni Pizza Dip

PREP: 10 MINUTES • COOK TIME: 10 MINUTES • TOTAL: 20 MINUTES • SERVES: 6

Ingredients

6 ounces cream cheese, softened
¾ cup shredded Italian cheese blend
¼ cup sour cream
1½ teaspoons dried Italian seasoning
¼ teaspoon garlic salt

¼ teaspoon onion powder
¾ cup pizza sauce
½ cup sliced miniature pepperoni
¼ cup sliced black olives
1 tablespoon thinly sliced green onion

Instructions

1. Cut-up raw vegetables, toasted baguette slices, pita chips, or tortilla chips, for serving. In a small bowl, combine the cream cheese, ¼ cup of the shredded cheese, the sour cream, Italian seasoning, garlic salt, and onion powder. Stir until smooth and the ingredients are well blended. Spread the mixture in a 6 × 3-inch round heatproof pan. Top with the pizza sauce, spreading to the edges. Sprinkle with the remaining ½ cup shredded cheese. Arrange the pepperoni slices on top of the cheese. Top with the black olives and green onion. Place the pan in the air fryer basket.

2. Set the Innsky air fryer to 350°F for 10 minutes, or until the pepperoni is beginning to brown on the edges and the cheese is bubbly and lightly browned. Let stand for 5 minutes before serving with vegetables, toasted baguette slices, pita chips, or tortilla chips.

Healthy Carrot Chips

PREP: 5 MINUTES • COOK TIME: 8 MINUTES • TOTAL: 13 MINUTES • SERVES: 6

Ingredients

1 pound carrots, peeled and sliced ⅛ inch thick
2 tablespoons olive oil

1 teaspoon sea salt

Instructions

1. In a large mixing bowl, combine the carrots, olive oil, and salt. Toss them together until the carrot slices are thoroughly coated with oil. Place the carrot chips in the Innsky air fryer basket in a single layer. (You may have to bake the carrot chips in more than one batch.)

2. Set the temperature of your Innsky AF to 360°F. Set the timer and bake for 3 minutes. Remove the air fryer drawer and shake to redistribute the chips for even cooking. Reset the timer and bake for 3 minutes more. Check the carrot chips for doneness. If you like them extra crispy, give the basket another shake and cook them for another 1 to 2 minutes. When the chips are done, release the Innsky air fryer basket from the drawer, pour the chips into a bowl, and serve.

Salt and Vinegar Pork Belly Chips

PREP: 5 MINUTES PLUS 30 MINUTES TO MARINATE • COOK TIME: 12 MINUTES • TOTAL: 47 MINUTES • SERVES: 4

Ingredients

1 pound slab pork belly
½ cup apple cider vinegar
Fine sea salt

FOR SERVING (OPTIONAL):
Guacamole
Pico de gallo

Instructions

1. Slice the pork belly into ⅛-inch-thick strips and place them in a shallow dish. Pour in the vinegar and stir to coat the pork belly. Place in the fridge to marinate for 30 minutes. Spray the Innsky air fryer basket with avocado oil. Preheat the Innsky air fryer to 400°F.

2. Remove the pork belly from the vinegar and place the strips in the Innsky air fryer basket in a single layer, leaving space between them. Cook in the air fryer for 10 to 12 minutes, until crispy, flipping after 5 minutes. Remove from the air fryer and sprinkle with salt. Serve with guacamole and pico de gallo, if desired. Served fresh.

Per serving: Calories 240; Fat 21g; Protein 13g; Total carbs 0g; Fiber 0g

Air Fried Homemade Potato Chips

PREP: 5 MINUTES • COOK TIME: 25 MINUTES • TOTAL: 30 MINUTES • SERVES: 4

Ingredients

4 yellow potatoes

1 tablespoon olive oil

1 tablespoon salt

Instructions

1. Using a mandoline or sharp knife, slice the potatoes into ⅛-inch-thick slices. In a medium mixing bowl, toss the potato slices with the olive oil and salt until the potatoes are thoroughly coated with oil. Place the potatoes in the Innsky air fryer basket in a single layer.

2. Set the temperature of your Innsky AF to 375°F. Set the timer and fry for 15 minutes.
   Shake the basket several times during cooking, so the chips crisp evenly and don't burn.
   Check to see if they are fork-tender; if not, add another 5 to 10 minutes, checking frequently. They will crisp up after they are removed from the air fryer. Season with additional salt, if desired.

Savory Potato Patties

PREP: 5 MINUTES PLUS 10 MINUTES TO STAND • COOK TIME: 10 MINUTES • TOTAL: 25 MINUTES • SERVES: 4

Ingredients

⅔ cup instant potato flakes

¼ cup frozen peas and carrots, thawed

2 tablespoons chopped fresh cilantro

1 tablespoon vegetable oil

½ teaspoon ground turmeric

½ teaspoon cumin seeds

¼ teaspoon ground cumin

½ teaspoon kosher salt

¼ to ½ teaspoon cayenne pepper

⅔ cup hot water

Vegetable oil spray

Instructions

1. In a medium bowl, combine the potato flakes, peas and carrots, cilantro, oil, turmeric, cumin seeds, ground cumin, salt, and cayenne. Add the hot water and stir gently until the ingredients are well combined. Cover and let stand for 10 minutes. Form the dough into 12 round, flat patties with even edges. Spray the Innsky air fryer basket with vegetable oil spray. Arrange half of the patties in the air fryer basket. Set the Innsky air fryer to 400°F for 10 minutes. After 5 minutes of cooking time, spray the patties with oil spray. Remove from basket with a flexible spatula. Repeat to cook the remaining patties, checking at 8 minutes for doneness. Serve hot.

Parmesan Dill Fried Pickles

PREP: 5 MINUTES • COOK TIME: 4 MINUTES • TOTAL: 9 MINUTES • SERVES: 4

Ingredients

1 (16-ounce) jar sliced dill pickles
⅔ cup panko bread crumbs
⅓ cup grated Parmesan cheese

¼ teaspoon dried dill
2 large eggs

Instructions

1. Line a platter with a double thickness of paper towels. Spread the pickles out in a single layer on the paper towels. Let the pickles drain on the towels for 20 minutes. After 20 minutes, pat the pickles again with a clean paper towel to get them as dry as possible before breading. Spray the Innsky air fryer basket with olive oil. In a small mixing bowl, combine the panko bread crumbs, Parmesan cheese, and dried dill. Mix well. In a separate small bowl, crack the eggs and beat until frothy. Dip each pickle into the egg mixture, then into the bread crumb mixture. Make sure the pickle is fully coated in breading. Place the breaded pickle slices in the greased Innsky air fryer basket in a single layer. Spray the pickles with a generous amount of olive oil.

2. Set the temperature of your Innsky AF to 390°F. Set the timer and fry for 4 minutes. Open the air fryer drawer and use tongs to flip the pickles. Spray them again with olive oil. Reset the timer and fry for another 4 minutes. Using tongs, remove the pickles from the drawer. Plate, serve, and enjoy!

Crispy Prosciutto-Wrapped Onion Rings

PREP: 10 MINUTES • COOK TIME: 10 MINUTES • TOTAL: 20 MINUTES • SERVES: ABOUT 1½ CUPS

Ingredients

3 large sweet onions
24 slices prosciutto or beef bacon
1 cup Ranch Dressing, for serving (optional; use dairy-free if needed)
Ranch Dressing
1 (8-ounce) package cream cheese softened
½ cup chicken or beef broth

½ teaspoon dried chives
½ teaspoon dried dill weed
½ teaspoon dried parsley
¼ teaspoon garlic powder
¼ teaspoon onion powder
⅛ teaspoon fine sea salt
⅛ teaspoon ground black pepper

Instructions

1. In a blender or using a hand mixer with a large bowl, mix together all the ingredients until well combined. Cover and refrigerate for 2 hours before serving (it will thicken up as it rests). Spray the Innsky air fryer basket with avocado oil. Preheat the Innsky air fryer to 400°F. Cut the onions into ⅔-inch-thick slices. Reserve the small inside rings for other recipes; you want the large rings, which are easy to wrap in prosciutto.
   Wrap each onion ring tightly in prosciutto.

2. Place the wrapped onion rings into the air fryer and cook for 10 minutes, or until the prosciutto is crispy. Serve with ranch dressing, if desired.
   Store leftovers in an airtight container in the fridge for up to 5 days. Reheat in a preheated 400°F air fryer for 5 minutes, or until heated through, flipping halfway through.

Per serving: Calories 318; Fat 28g; Protein 14g; Total carbs 3g; Fiber 0.4g

Loaded Potato Skins

PREP: 10 MINUTES • COOK TIME: 12 MINUTES • TOTAL: 22 MINUTES • SERVES: 4

Ingredients

4 medium russet potatoes, baked
Olive oil
Salt
Freshly ground black pepper
2 cups shredded Cheddar cheese

4 slices cooked bacon, chopped
Finely chopped scallions, for topping
Sour cream, for topping
Finely chopped olives, for topping

Instructions

1. Spray the Innsky air fryer basket with oil. Cut each baked potato in half. Using a large spoon, scoop out the center of each potato half, leaving about 1 inch of the potato flesh around the edges and the bottom.
Rub olive oil over the inside of each baked potato half and season with salt and pepper, then place the potato skins in the greased air fryer basket.
2. Set the temperature of your Innsky AF to 400°F. Set the timer and bake for 10 minutes.
After 10 minutes, remove the potato skins and fill them with the shredded Cheddar cheese and bacon, then bake in the air fryer for another 2 minutes, just until the cheese is melted. Garnish the potato skins with the scallions, sour cream, and olives.

Per Serving: Calories: 487; Fat: 31g; Saturated fat: 16g; Carbohydrate: 29g; Fiber: 5g; Sugar: 1g; Protein: 24g; Iron: 5mg; Sodium: 986mg

Bacon-Wrapped Asparagus

PREP: 5 MINUTES • COOK TIME: 10 MINUTES • TOTAL: 15 MINUTES • SERVES: 4

Ingredients

1 pound asparagus, trimmed (about 24 spears)
4 slices bacon or beef bacon

½ cup Ranch Dressing , for serving
2 tablespoons chopped fresh chives, for garnish

Instructions

1. Spray the Innsky air fryer basket with avocado oil. Preheat the Innsky air fryer to 400°F. Slice the bacon down the middle, making long, thin strips. Wrap 1 slice of bacon around 3 asparagus spears and secure each end with a toothpick. Repeat with the remaining bacon and asparagus. Place the asparagus bundles in the air fryer in a single layer. (If you're using a smaller air fryer, cook in batches if necessary.)
2. Cook for 8 minutes for thin stalks, 10 minutes for medium to thick stalks, or until the asparagus is slightly charred on the ends and the bacon is crispy. Serve with ranch dressing and garnish with chives. Best served fresh. Store leftovers in an airtight container in the fridge for up to 5 days.

Per serving: Calories 241; Fat 22g; Protein 7g; Total carbs 6g; Fiber 3g

Fresh Homemade Potato Wedges

PREP: 5 MINUTES • COOK TIME: 25 MINUTES • TOTAL: 30 MINUTES • SERVES: 4

Ingredients

4 russet potatoes
2 teaspoons salt, divided
1 teaspoon freshly ground black pepper

1 teaspoon paprika
1 to 3 tablespoons olive oil, divided

Instructions

1. Cut the potatoes into ½-inch-thick wedges. Try to make the wedges uniform in size, so they cook at an even rate. In a medium mixing bowl, combine the potato wedges with 1 teaspoon of salt, pepper, paprika, and 1 tablespoon of olive oil. Toss until all the potatoes are thoroughly coated with oil. Add additional oil, if needed. Place the potato wedges in the Innsky air fryer basket in a single layer.

2. Set the temperature of your Innsky AF to 400°F. Set the timer and roast for 5 minutes.
   After 5 minutes, remove the air fryer drawer and shake the potatoes to keep them from sticking. Reset the timer and roast the potatoes for another 5 minutes, then shake again. Repeat this process until the potatoes have cooked for a total of 20 minutes.
   Check and see if the potatoes are cooked. If they are not fork-tender, roast for 5 minutes more. Using tongs, remove the potato wedges from the Innsky air fryer basket and transfer them to a bowl. Toss with the remaining salt.

Per Serving: Calories: 210; Fat: 7g; Saturated fat: 1g; Carbohydrate: 34g; Fiber: 6g; Sugar: 3g; Protein: 4g; Iron: 1mg; Sodium: 1176mg

Reuben Egg Rolls

PREP: 20 MINUTES • COOK TIME: 30 MINUTES • TOTAL: 50 MINUTES • SERVES: 20 EGG ROLLS

Ingredients

1 (8-ounce) package cream cheese, softened
½ pound cooked corned beef, chopped
½ cup drained and chopped sauerkraut
½ cup shredded Swiss cheese (about 2 ounces)
20 slices prosciutto
THOUSAND ISLAND DIPPING SAUCE:
¾ cup mayonnaise
¼ cup chopped dill pickles

¼ cup tomato sauce
2 tablespoons Swerve confectioners'-style sweetener or equivalent amount of liquid or powdered sweetener
⅛ teaspoon fine sea salt
Fresh thyme leaves, for garnish
Ground black pepper, for garnish
Sauerkraut, for serving (optional)

Instructions

1. Spray the Innsky air fryer basket with avocado oil. Preheat the Innsky air fryer to 400°F.

    Make the filling: Place the cream cheese in a medium-sized bowl and stir to break it up. Add the corned beef, sauerkraut, and Swiss cheese and stir well to combine.

    Assemble the egg rolls: Lay 1 slice of prosciutto on a sushi mat or a sheet of parchment paper with a short end toward you. Lay another slice of prosciutto on top of it at a right angle, forming a cross. Spoon 3 to 4 tablespoons of the filling into the center of the cross.

    Fold the sides of the top slice up and over the filling to form the ends of the roll. Tightly roll up the long piece of prosciutto, starting at the edge closest to you, into a tight egg roll shape that overlaps by an inch or so. Repeat with the remaining prosciutto and filling.

    Place the egg rolls in the air fryer seam side down, leaving space between them.

2. Cook for 10 minutes, or until the outside is crispy.

    While the egg rolls are cooking, make the dipping sauce: In a small bowl, combine the mayo, pickles, tomato sauce, sweetener, and salt. Stir well and garnish with thyme and ground black pepper. (The dipping sauce can be made up to 3 days ahead.)

    Serve the egg rolls with the dipping sauce and sauerkraut if desired. Best served fresh. Store leftovers in an airtight container in the refrigerator for up to 5 days or in the freezer for up to a month. Reheat in a preheated 400°F air fryer for 4 minutes, or until heated through and crispy.

Per serving: Calories 321; Fat 29g; Protein 13g; Total carbs 1g; Fiber 0.1g

Smoky Salmon Dip

PREP: 10 MINUTES • COOK TIME: 7 MINUTES • TOTAL: 17 MINUTES • SERVES: 6

Ingredients

1 (6-ounce) can boneless, skinless salmon
8 ounces cream cheese, softened
1 tablespoon liquid smoke (optional)
⅓ cup chopped pecans
½ cup chopped green onions

1 teaspoon kosher salt
1 to 2 teaspoons black pepper
¼ teaspoon smoked paprika, for garnish
Cucumber and celery slices, cocktail rye bread, and/or crackers

Instructions

1. In a 6 × 3-inch round heatproof pan, combine the salmon, softened cream cheese, liquid smoke (if using), pecans, ¼ cup of the green onions, and the salt and pepper. Stir until well combined. Place the pan in the air fryer basket.

2. Set the Innsky air fryer to 400°F for 7 minutes, or until the cheese melts.
   Sprinkle with the paprika and top with the remaining ¼ cup green onions. Serve with sliced vegetables, cocktail breads, and/or crackers.

Mozzarella Sticks

PREP:15 MINUTES PLUS 2 HOURS TO FREEZE•COOK TIME:14 MINUTES•TOTAL:2 HOURS 29 MINUTES •SERVES: 24

Ingredients

DOUGH:
1¾ cups shredded mozzarella cheese (about 7 ounces)
2 tablespoons unsalted butter
1 large egg, beaten
¾ cup blanched almond flour
⅛ teaspoon fine sea salt
24 pieces of string cheese

SPICE MIX:
¼ cup grated Parmesan cheese
3 tablespoons garlic powder
1 tablespoon dried oregano leaves
1 tablespoon onion powder
FOR SERVING (OPTIONAL):
½ cup marinara sauce
½ cup pesto

Instructions

1. Make the dough: Place the mozzarella and butter in a large microwave-safe bowl and microwave for 1 to 2 minutes, until the cheese is entirely melted. Stir well.
   Add the egg and, using a hand mixer on low, combine well. Add the almond flour and salt and combine well with the mixer. Lay a piece of parchment paper on the countertop and place the dough on it. Knead it for about 3 minutes; the dough should be thick yet pliable. Scoop up 3 tablespoons of the dough and flatten it into a very thin 3½ by 2-inch rectangle. Place one piece of string cheese in the center and use your hands to press the dough tightly around it. Repeat with the remaining string cheese and dough. In a shallow dish, combine the spice mix ingredients. Place a wrapped piece of string cheese in the dish and roll while pressing down to form a nice crust. Repeat with the remaining pieces of string cheese. Place in the freezer for 2 hours. Ten minutes before air frying, spray the Innsky air fryer basket with avocado oil and Preheat the Innsky air fryer to 425°F.

2. Place the frozen mozzarella sticks in the air fryer basket, leaving space between them, and cook for 9 to 12 minutes, until golden brown. Remove from the air fryer and serve with marinara sauce and pesto, if desired. Store leftovers in an airtight container in the refrigerator for up to 3 days or in the freezer for up to a month. Reheat in a preheated 425°F air fryer for 4 minutes, or until warmed through.

Per serving: Calories 337; Fat 27g; Protein 23g; Total carbs 4g; Fiber 1g

Smoky Eggplant Tahini Dip

PREP: 20 MINUTES • COOK TIME: 15 MINUTES • TOTAL: 35 MINUTES • SERVES: 6

Ingredients

1 large eggplant
2 tablespoons olive oil
5 cloves garlic, minced
2 tablespoons tahini (sesame paste)
½ teaspoon kosher salt

1 tablespoon extra-virgin olive oil
½ teaspoon smoked paprika
2 tablespoons chopped fresh parsley
Raw vegetables and/or pita bread, for serving

Instructions

1 Rub the eggplant all over with the oil and place in the air fryer basket.

2 Set the Innsky air fryer to 400°F for 15 minutes, or until the eggplant's skin is well browned. Place the eggplant in a bowl, cover with foil, and let steam for 10 minutes to finish cooking. Holding the eggplant over the bowl, remove the skin and discard. Mash the eggplant along with the juices. Add the garlic, tahini, and salt and mix well. Scrape the dip into a bowl. Create a well in the dip using the back of a spoon. Pour the olive oil into the well. Top with paprika and chopped parsley. Serve with vegetables or pita bread.

Air Fryer Stuffed Mushrooms

PREP: 5 MINUTES • COOK TIME: 10 MINUTES • TOTAL: 15 MINUTES • SERVES: 4

Ingredients

12 medium button mushrooms
½ cup bread crumbs
1 teaspoon salt

½ teaspoon freshly ground black pepper
5 to 6 tablespoons olive oil

Instructions

1 Spray the Innsky air fryer basket with olive oil. Separate the cap from the stem of each mushroom. Discard the stems. In a small mixing bowl, combine the bread crumbs, salt, pepper, and olive oil until you have a wet mixture. Rub the mushrooms with olive oil on all sides. Using a spoon, fill each mushroom with the bread crumb stuffing. Place the mushrooms in the greased Innsky air fryer basket in a single layer.

2 Set the temperature of your Innsky AF to 360°F. Set the timer and bake for 10 minutes. Using tongs, remove the mushrooms from the air fryer, place them on a platter, and serve.

Crispy Calamari Rings

PREP: 10 MINUTES • COOK TIME: 15 MINUTES • TOTAL: 25 MINUTES • SERVES: 4

Ingredients

2 large egg yolks
1 cup powdered Parmesan cheese
¼ cup coconut flour
3 teaspoons dried oregano leaves
½ teaspoon garlic powder

½ teaspoon onion powder
1 pound calamari, sliced into rings
Fresh oregano leaves, for garnish (optional)
1 cup marinara sauce, for serving (optional)
Lemon slices, for serving (optional)

Instructions

1 Spray the Innsky air fryer basket with avocado oil. Preheat the Innsky air fryer to 400°F. In a shallow dish, whisk the egg yolks. In a separate bowl, mix together the Parmesan, coconut flour, and spices. Dip the calamari rings in the egg yolks, tap off any excess egg, then dip them into the cheese mixture and coat well. Use your hands to press the coating onto the calamari if necessary. Spray the coated rings with avocado oil.

2 Place the calamari rings in the air fryer, leaving space between them, and cook for 15 minutes, or until golden brown. Garnish with fresh oregano, if desired, and serve with marinara sauce for dipping and lemon slices, if desired.

Per serving: Calories 287; Fat 13g; Protein 28g; Total carbs 11g; Fiber 3g

Air Fryer Bacon-Wrapped Jalapeño Poppers
PREP: 5 MINUTES • COOK TIME: 12 MINUTES • TOTAL: 17 MINUTES • SERVES: 12

Ingredients
12 jalapeño peppers
1 (8-ounce) package cream cheese, at room temperature
1 cup shredded Cheddar cheese

1 teaspoon onion powder
1 teaspoon salt
½ teaspoon freshly ground black pepper
12 slices bacon, cut in half

Instructions
1   Spray the Innsky air fryer basket with olive oil. Cut each pepper in half, then use a spoon to scrape out the veins and seeds. In a small mixing bowl, mix together the cream cheese, Cheddar cheese, onion powder, salt, and pepper. Using a small spoon, fill each pepper half with the cheese mixture. Wrap each stuffed pepper half with a half slice of bacon. Place the bacon-wrapped peppers into the greased Innsky air fryer basket in a single layer.
2   Set the temperature of your Innsky AF for 320°F. Set the timer and bake for 12 minutes. Using tongs, remove the peppers from the air fryer, place them on a platter, and serve.

Bloomin' Onion
PREP: 10 MINUTES • COOK TIME: 35 MINUTES • TOTAL: 45 MINUTES • SERVES: 8

Ingredients
1 extra-large onion (about 3 inches in diameter)
2 large eggs
1 tablespoon water
½ cup powdered Parmesan cheese
2 teaspoons paprika
1 teaspoon garlic powder
¼ teaspoon cayenne pepper
¼ teaspoon fine sea salt

¼ teaspoon ground black pepper
FOR GARNISH (OPTIONAL):
Fresh parsley leaves
Powdered Parmesan cheese
FOR SERVING (OPTIONAL):
Prepared yellow mustard
Ranch Dressing
Reduced-sugar or sugar-free ketchup

Instructions
1   Spray the Innsky air fryer basket with avocado oil. Preheat the Innsky air fryer to 350°F.
   Using a sharp knife, cut the top ½ inch off the onion and peel off the outer layer. Cut the onion into 8 equal sections, stopping 1 inch from the bottom, you want the onion to stay together at the base. Gently spread the sections, or "petals," apart. Crack the eggs into a large bowl, add the water, and whisk well. Place the onion in the dish and coat it well in the egg. Use a spoon to coat the inside of the onion and all of the petals. In a small bowl, combine the Parmesan, seasonings, salt, and pepper.
   Place the onion in a 6-inch pie pan or casserole dish. Sprinkle the seasoning mixture all over the onion and use your fingers to press it into the petals. Spray the onion with avocado oil. Loosely cover the onion with parchment paper and then foil.
2   Place the dish in the air fryer. Cook for 30 minutes, then remove it from the air fryer and increase the air fryer temperature to 400°F. Remove the foil and parchment and spray the onion with avocado oil again. Protecting your hands with oven-safe gloves or a tea towel, transfer the onion to the air fryer basket. Cook for an additional 3 to 5 minutes, until light brown and crispy.
   Garnish with fresh parsley and powdered Parmesan, if desired. Serve with mustard, ranch dressing, and ketchup, if desired.

Per serving: Calories 51; Fat 3g; Protein 4g; Total carbs 3g; Fiber 0.4g

Prosciutto-Wrapped Guacamole Rings

PREP:10MINUTES PLUS 2 HOURS TO FREEZE•COOK TIME: 6 MINUTES•TOTAL:2HOURS 16 MINUTES•SERVES:8

Ingredients

GUACAMOLE:

2 avocados, halved, pitted, and peeled

3 tablespoons lime juice, plus more to taste

2 small plum tomatoes, diced

½ cup finely diced onions

2 small cloves garlic, smashed to a paste

3 tablespoons chopped fresh cilantro leaves

½ scant teaspoon fine sea salt

½ scant teaspoon ground cumin

3 small onions (about 1½ inches in diameter), cut into ½-inch-thick slices

8 slices prosciutto

Instructions

1   Make the guacamole: Place the avocados and lime juice in a large bowl and mash with a fork until it reaches your desired consistency. Add the tomatoes, onions, garlic, cilantro, salt, and cumin and stir until well combined. Taste and add more lime juice if desired. Set aside half of the guacamole for serving. Place a piece of parchment paper on a tray that fits in your freezer and place the onion slices on it, breaking the slices apart into 8 rings. Fill each ring with about 2 tablespoons of guacamole. Place the tray in the freezer for 2 hours. Spray the Innsky air fryer basket with avocado oil. Preheat the Innsky air fryer to 400°F.Remove the rings from the freezer and wrap each in a slice of prosciutto.

2

Place them in the air fryer basket, leaving space between them, and cook for 6 minutes, flipping halfway through. Use a spatula to remove the rings from the air fryer. Serve with the reserved half of the guacamole.

Prosciutto Pierogi

PREP: 15 MINUTES • COOK TIME: 20 MINUTES • TOTAL: 35 MINUTES • SERVES: 4

Ingredients

1 cup chopped cauliflower

2 tablespoons diced onions

1 tablespoon unsalted butter melted

Pinch of fine sea salt

½ cup shredded sharp cheddar cheese

8 slices prosciutto

Fresh oregano leaves, for garnish (optional)

Instructions

1   Preheat the Innsky air fryer to 350°F. Lightly grease a 7-inch pie pan or a casserole dish that will fit in your air fryer.

Make the filling: Place the cauliflower and onion in the pan. Drizzle with the melted butter and sprinkle with the salt. Using your hands, mix everything together, making sure the cauliflower is coated in the butter.

Place the cauliflower mixture in the air fryer and cook for 10 minutes, until fork-tender, stirring halfway through.

Transfer the cauliflower mixture to a food processor or high-powered blender. Spray the Innsky air fryer basket with avocado oil and increase the air fryer temperature to 400°F.

Pulse the cauliflower mixture in the food processor until smooth. Stir in the cheese.

Assemble the pierogi: Lay 1 slice of prosciutto on a sheet of parchment paper with a short end toward you. Lay another slice of prosciutto on top of it at a right angle, forming a cross. Spoon about 2 heaping tablespoons of the filling into the center of the cross.

Fold each arm of the prosciutto cross over the filling to form a square, making sure that the filling is well covered. Using your fingers, press down around the filling to even out the square shape. Repeat with the rest of the prosciutto and filling.

2   Spray the pierogi with avocado oil and place them in the air fryer basket. Cook for 10 minutes, or until crispy. Garnish with oregano before serving, if desired. Store leftovers in an airtight container in the fridge for up to 4 days. Reheat in a preheated 400°F air fryer for 3 minutes, or until heated through.

Per serving: Calories **150**; Fat **11g**; Protein **11g**; Total carbs **2g**; Fiber **1g**

Vegetables Recipes

Air Fryer Asparagus
PREP: 5 MINUTES • COOK TIME: 8 MINUTES • TOTAL: 13 MINUTES • SERVES: 2

Ingredients

Nutritional yeast

Olive oil non-stick spray

One bunch of asparagus

Instructions

Wash asparagus and then trim off thick, woody ends. Spray asparagus with olive oil spray and sprinkle with yeast. In your Air fryer, lay asparagus in a singular layer. Set the temperature of your Innsky AF to 360°F, and set time to 8 minutes.

Simple Roasted Garlic Asparagus
PREP: 5 MINUTES • COOK TIME: 10 MINUTES • TOTAL: 15 MINUTES • SERVES: 4

Ingredients

1 pound asparagus

2 tablespoons olive oil

1 tablespoon balsamic vinegar

2 teaspoons minced garlic

Salt

Freshly ground black pepper

Instructions

1. Cut or snap off the white end of the asparagus.
    In a large bowl, combine the asparagus, olive oil, vinegar, garlic, salt, and pepper.
    Using your hands, gently mix all the ingredients together, making sure that the asparagus is thoroughly coated.
    Lay out the asparagus in the Innsky air fryer basket or on an air fryer–size baking sheet set in the basket.
2. Set the temperature of your Innsky AF to 400°F. Set the timer and roast for 5 minutes.
    Using tongs, flip the asparagus. Reset the timer and roast for 5 minutes more.

Per Serving: Calories: 86; Fat: 7g; Saturated fat: 1g; Carbohydrate: 5g; Fiber: 2g; Sugar: 2g; Protein: 3g; Iron: 2mg; Sodium: 41mg

Ranch Kale Chips
PREP: 5 MINUTES • COOK TIME: 10 MINUTES • TOTAL: 15 MINUTES • SERVES: 8 CUPS

Ingredients

½ teaspoon dried chives

½ teaspoon dried dill weed

½ teaspoon dried parsley

¼ teaspoon garlic powder

¼ teaspoon onion powder

⅛ teaspoon fine sea salt

⅛ teaspoon ground black pepper

2 large bunches kale

Instructions

1. Spray the Innsky air fryer basket with avocado oil. Preheat the Innsky air fryer to 360°F.
    Place the seasonings, salt, and pepper in a small bowl and mix well. Wash the kale and pat completely dry. Use a sharp knife to carve out the thick inner stems, then spray the leaves with avocado oil and sprinkle them with the seasoning mix.
2. Place the kale leaves in the air fryer in a single layer and cook for 10 minutes, shaking and rotating the chips halfway through. Transfer the baked chips to a baking sheet to cool completely and crisp up. Repeat with the remaining kale. Sprinkle the cooled chips with salt before serving, if desired.
    Kale chips can be stored in an airtight container at room temperature for up to 1 week, but they are best eaten within 3 days.

Per serving: Calories 11; Fat 0.2g; Protein 1g; Total carbs 2g; Fiber 0.4g

Almond Flour Battered And Crisped Onion Rings
PREP: 5 MINUTES • COOK TIME: 15 MINUTES • TOTAL: 20 MINUTES • SERVES: 3

Ingredients
½ cup almond flour
¾ cup coconut milk
1 big white onion, sliced into rings
1 egg, beaten

1 tablespoon baking powder
1 tablespoon smoked paprika
Salt and pepper to taste

Instructions
1. Preheat the Innsky air fryer for 5 minutes. In a mixing bowl, mix the almond flour, baking powder, smoked paprika, salt and pepper. In another bowl, combine the eggs and coconut milk. Soak the onion slices into the egg mixture.
    Dredge the onion slices in the almond flour mixture.
2. Place in the air fryer basket. Close and cook for 15 minutes at 325 °F.
    Halfway through the cooking time, shake the fryer basket for even cooking.

Per Serving: Calories: 217; Fat: 17.9g; Protein: 5.3g

Crispy Nacho Avocado Fries
PREP: 10 MINUTES • COOK TIME: 15 MINUTES • TOTAL: 25 MINUTES • SERVES: 6

Ingredients
3 firm, barely ripe avocados, halved, peeled, and pitted
2 cups pork dust
2 teaspoons fine sea salt
2 teaspoons ground black pepper
2 teaspoons ground cumin
1 teaspoon chili powder

1 teaspoon paprika
½ teaspoon garlic powder
½ teaspoon onion powder
2 large eggs
Salsa, for serving (optional)
Fresh chopped cilantro leaves, for garnish (optional)

Instructions
1. Spray the Innsky air fryer basket with avocado oil. Preheat the Innsky air fryer to 400°F.
    Slice the avocados into thick-cut french fry shapes. In a bowl, mix together the pork dust, salt, pepper, and seasonings. In a separate shallow bowl, beat the eggs.
    Dip the avocado fries into the beaten eggs and shake off any excess, then dip them into the pork dust mixture. Use your hands to press the breading into each fry.
    Spray the fries with avocado oil and place them in the Innsky air fryer basket in a single layer, leaving space between them. If there are too many fries to fit in a single layer, work in batches.
2.
    Cook in the air fryer for 13 to 15 minutes, until golden brown, flipping after 5 minutes.
    Serve with salsa, if desired, and garnish with fresh chopped cilantro, if desired. Best served fresh. Store leftovers in an airtight container in the fridge for up to 5 days. Reheat in a preheated 400°F air fryer for 3 minutes, or until heated through.

Per serving: Calories 282; Fat 22g; Protein 15g; Total carbs 9g; Fiber 7g

Chermoula-Roasted Beets

PREP: 15 MINUTES • COOK TIME: 25 MINUTES • TOTAL: 40 MINUTES • SERVES: 4

Ingredients

For the Chermoula
1 cup packed fresh cilantro leaves
½ cup packed fresh parsley leaves
6 cloves garlic, peeled
2 teaspoons smoked paprika
2 teaspoons ground cumin
1 teaspoon ground coriander
½ to 1 teaspoon cayenne pepper

Pinch crushed saffron (optional)
½ cup extra-virgin olive oil
Kosher salt
For the Beets
3 medium beets, trimmed, peeled, and cut into 1-inch chunks
2 tablespoons chopped fresh cilantro
2 tablespoons chopped fresh parsley

Instructions

1. *For the chermoula*: In a food processor, combine the cilantro, parsley, garlic, paprika, cumin, coriander, and cayenne. Pulse until coarsely chopped. Add the saffron, if using, and process until combined. With the food processor running, slowly add the olive oil in a steady stream; process until the sauce is uniform. Season to taste with salt.
   For the beets: In a large bowl, drizzle the beets with ½ cup of the chermoula, or enough to coat.
2. Arrange the beets in the air fryer basket. Set the Innsky air fryer to 375°F for 25 to minutes, or until the beets are tender. Transfer the beets to a serving platter. Sprinkle with chopped cilantro and parsley and serve.

Jalapeño Poppers

PREP: 10 MINUTES • COOK TIME: 10 MINUTES • TOTAL: 20 MINUTES • SERVES: 4

Ingredients

12-18 whole fresh jalapeño
1 cup nonfat refried beans
1 cup shredded Monterey Jack or extra-sharp cheddar cheese
1 scallion, sliced

1 teaspoon salt, divided
1/4 cup all-purpose flour
2 large eggs
1/2 cup fine cornmeal
Olive oil or canola oil cooking spray

Instructions

1. Start by slicing each jalapeño lengthwise on one side. Place the jalapeños side by side in a microwave safe bowl and microwave them until they are slightly soft; usually around 5 minutes. While your jalapeños cook; mix refried beans, scallions, 1/2 teaspoon salt, and cheese in a bowl. Once your jalapeños are softened you can scoop out the seeds and add one tablespoon of your refried bean mixture. Press the jalapeño closed around the filling. Beat your eggs in a small bowl and place your flour in a separate bowl. In a third bowl mix your cornmeal and the remaining salt in a third bowl. Roll each pepper in the flour, then dip it in the egg, and finally roll it in the cornmeal making sure to coat the entire pepper. Place the peppers on a flat surface and coat them with a cooking spray; olive oil cooking spray is suggested.
2. Cook in your Air fryer at 400 degrees for 5 minutes, turn each pepper, and then cook for another 5 minutes; serve hot.

Fried Plantains

PREP: 10 MINUTES • COOK TIME: 8 MINUTES • TOTAL: 18 MINUTES • SERVES: 2

Ingredients

2 ripe plantains, peeled and cut at a diagonal into ½-inch-thick pieces

3 tablespoons ghee, melted
¼ teaspoon kosher salt

Instructions

1. In a medium bowl, toss the plantains with the ghee and salt.
2. Arrange the plantain pieces in the air fryer basket. Set the Innsky air fryer to 400°F for 8 minutes. The plantains are done when they are soft and tender on the inside, and have plenty of crisp, sweet, brown spots on the outside.

Bacon-Wrapped Asparagus

PREP: 5 MINUTES • COOK TIME: 10 MINUTES • TOTAL: 15 MINUTES • SERVES: 4

Ingredients

1 pound asparagus, trimmed (about 24 spears)
4 slices bacon or beef bacon
½ cup Ranch Dressin for serving

3 tablespoons chopped fresh chives, for garnish

Instructions

1. Spray the Innsky air fryer basket with avocado oil. Preheat the Innsky air fryer to 400°F.
   Slice the bacon down the middle, making long, thin strips. Wrap 1 slice of bacon around 3 asparagus spears and secure each end with a toothpick. Repeat with the remaining bacon and asparagus.
2. Place the asparagus bundles in the air fryer in a single layer. (If you're using a smaller air fryer, cook in batches if necessary.) Cook for 8 minutes for thin stalks, 10 minutes for medium to thick stalks, or until the asparagus is slightly charred on the ends and the bacon is crispy.
   Serve with ranch dressing and garnish with chives. Best served fresh.

Per serving: Calories 241; Fat 22g; Protein 7g; Total carbs 6g; Fiber 3g

Air Fried Roasted Corn on The Cob

PREP: 5 MINUTES • COOK TIME: 10 MINUTES • TOTAL: 15 MINUTES • SERVES: 4

Ingredients

1 tablespoon vegetable oil
4 ears of corn, husks and silk removed
Unsalted butter, for topping

Salt, for topping
Freshly ground black pepper, for topping

Instructions

1. Rub the vegetable oil onto the corn, coating it thoroughly.
2. Set the temperature of your Innsky AF to 400°F. Set the timer and grill for 5 minutes.
   Using tongs, flip or rotate the corn.
   Reset the timer and grill for 5 minutes more.
   Serve with a pat of butter and a generous sprinkle of salt and pepper.

Per Serving: Calories: 265; Fat: 17g; Saturated fat: 8g; Carbohydrate: 29g; Fiber: 4g; Sugar: 5g; Protein: 5g; Iron: 4mg; Sodium: 252mg

Parmesan Breaded Zucchini Chips

PREP: 15 MINUTES • COOK TIME: 20 MINUTES • TOTAL: 35 MINUTES • SERVES: 5

Ingredients

For the zucchini chips:
2 medium zucchini
2 eggs
⅓ cup bread crumbs
⅓ cup grated Parmesan cheese
Salt
Pepper
Cooking oil

For the lemon aioli:
½ cup mayonnaise
½ tablespoon olive oil
Juice of ½ lemon
1 teaspoon minced garlic
Salt
Pepper

Instructions

1   To make the zucchini chips: Slice the zucchini into thin chips (about ⅛ inch thick) using a knife or mandoline. In a small bowl, beat the eggs. In another small bowl, combine the bread crumbs, Parmesan cheese, and salt and pepper to taste. Spray the Innsky air fryer basket with cooking oil. Dip the zucchini slices one at a time in the eggs and then the bread crumb mixture. You can also sprinkle the bread crumbs onto the zucchini slices with a spoon. Place the zucchini chips in the Air fryer basket, but do not stack.

2   Cook in batches. Spray the chips with cooking oil from a distance (otherwise, the breading may fly off). Cook for 10 minutes. Remove the cooked zucchini chips from the air fryer, then repeat step 5 with the remaining zucchini.

To make the lemon aioli: While the zucchini is cooking, combine the mayonnaise, olive oil, lemon juice, and garlic in a small bowl, adding salt and pepper to taste. Mix well until fully combined. Cool the zucchini and serve alongside the aioli.

Per Serving: Calories: 192; Fat: 13g; Protein: 6g; Fiber: 4g

Green Beans & Bacon

PREP: 10 MINUTES • COOK TIME: 20 MINUTES • TOTAL: 35 MINUTES • SERVES: 4

Ingredients

3 cups frozen cut green beans
1 medium onion, chopped
3 slices bacon, chopped

¼ cup water
Kosher salt and black pepper

Instructions

1. In a 6 × 3-inch round heatproof pan, combine the frozen green beans, onion, bacon, and water. Toss to combine. Place the pan in the air fryer basket.

2. Set the Innsky air fryer to 375°F for 15 minutes.
    Raise the air fryer temperature to 400°F for 5 minutes. Season the beans with salt and pepper to taste and toss well.
    Remove the pan from the Innsky air fryer basket and cover with foil. Let the beans rest for 5 minutes before serving.

Bell Pepper-Corn Wrapped in Tortilla

PREP: 5 MINUTES • COOK TIME: 15 MINUTES • TOTAL: 20 MINUTES • SERVES: 4

Ingredients

1 small red bell pepper, chopped
1 small yellow onion, diced
1 tablespoon water
2 cobs grilled corn kernels

4 large tortillas
4 pieces commercial vegan nuggets, chopped
mixed greens for garnish

Instructions

1. Preheat the Innsky air fryer to 400°F. In a skillet heated over medium heat, water sauté the vegan nuggets together with the onions, bell peppers, and corn kernels. Set aside.

   Place filling inside the corn tortillas.

2. Fold the tortillas and place inside the air fryer and cook for 15 minutes until the tortilla wraps are crispy. Serve with mix greens on top.

Bloomin' Onion

PREP: 10 MINUTES • COOK TIME: 35 MINUTES • TOTAL: 45 MINUTES • SERVES: 8

Ingredients

1 extra-large onion
2 large eggs
1 tablespoon water
½ cup powdered Parmesan cheese
2 teaspoons paprika
1 teaspoon garlic powder
¼ teaspoon cayenne pepper
¼ teaspoon fine sea salt

¼ teaspoon ground black pepper
FOR GARNISH (OPTIONAL):
Fresh parsley leaves
Powdered Parmesan cheese
FOR SERVING (OPTIONAL):
Prepared yellow mustard
Ranch Dressing
Reduced-sugar or sugar-free ketchup

Instructions

1. Spray the Innsky air fryer basket with avocado oil. Preheat the Innsky air fryer to 350°F.

   Using a sharp knife, cut the top ½ inch off the onion and peel off the outer layer. Cut the onion into 8 equal sections, stopping 1 inch from the bottom—you want the onion to stay together at the base. Gently spread the sections, or "petals," apart.

   Crack the eggs into a large bowl, add the water, and whisk well. Place the onion in the dish and coat it well in the egg. Use a spoon to coat the inside of the onion and all of the petals.

   In a small bowl, combine the Parmesan, seasonings, salt, and pepper.

   Place the onion in a 6-inch pie pan or casserole dish. Sprinkle the seasoning mixture all over the onion and use your fingers to press it into the petals. Spray the onion with avocado oil.

   Loosely cover the onion with parchment paper and then foil.

2. Place the dish in the air fryer. Cook for 30 minutes, then remove it from the air fryer and increase the air fryer temperature to 400°F.

   Remove the foil and parchment and spray the onion with avocado oil again. Protecting your hands with oven-safe gloves or a tea towel, transfer the onion to the air fryer basket. Cook for an additional 3 to 5 minutes, until light brown and crispy.

   Garnish with fresh parsley and powdered Parmesan, if desired. Serve with mustard, ranch dressing, and ketchup, if desired.

   Store leftovers in an airtight container in the fridge for up to 4 days. Reheat in a preheated 400°F air fryer for 3 to 5 minutes, until warm and crispy.

Per serving: Calories 51; Fat 3g; Protein 4g; Total carbs 3g; Fiber 0.4g

Mexican Corn In A Cup
PREP: 5 MINUTES • COOK TIME: 10 MINUTES • TOTAL: 15 MINUTES • SERVES: 4

Ingredients

4 cups (32-ounce bag) frozen corn kernels (do not thaw)
Vegetable oil spray
2 tablespoons butter
¼ cup sour cream
¼ cup mayonnaise

¼ cup grated Parmesan cheese (or feta, cotija, or queso fresco)
2 tablespoons fresh lemon or lime juice
1 teaspoon chili powder
Chopped fresh green onion (optional)
Chopped fresh cilantro (optional)

Instructions

1. Place the corn in the bottom of the Innsky air fryer basket and spray with vegetable oil spray.
2. Set the Innsky air fryer to 350°F for 10 minutes.
   Transfer the corn to a serving bowl. Add the butter and stir until melted. Add the sour cream, mayonnaise, cheese, lemon juice, and chili powder; stir until well combined. Serve immediately with green onion and cilantro (if using).

Air Fried Honey Roasted Carrots
PREP: 5 MINUTES • COOK TIME: 12 MINUTES • TOTAL: 17 MINUTES • SERVES: 4

Ingredients

3 cups baby carrots
1 tablespoon extra-virgin olive oil
1 tablespoon honey

Salt
Freshly ground black pepper
Fresh dill (optional)

Instructions

1. In a large bowl, combine the carrots, olive oil, honey, salt, and pepper. Make sure that the carrots are thoroughly coated with oil. Place the carrots in the air fryer basket.
2. Set the temperature of your Innsky AF to 390°F. Set the timer and roast for 12 minutes, or until fork-tender. Remove the air fryer drawer and release the air fryer basket. Pour the carrots into a bowl, sprinkle with dill, if desired, and serve.

Crispy Sesame-Ginger Broccoli
PREP: 10 MINUTES • COOK TIME: 15 MINUTES • TOTAL: 25 MINUTES • SERVES: 4

Ingredients

3 tablespoons toasted sesame oil
2 teaspoons sesame seeds
1 tablespoon chili-garlic sauce
2 teaspoons minced fresh ginger

½ teaspoon kosher salt
½ teaspoon black pepper
1 (16-ounce) package frozen broccoli florets (do not thaw)

Instructions

1. In a large bowl, combine the sesame oil, sesame seeds, chili-garlic sauce, ginger, salt, and pepper. Stir until well combined. Add the broccoli and toss until well coated.
2. Arrange the broccoli in the air fryer basket. Set the Innsky air fryer to 325°F for 15 minutes, or until the broccoli is crisp, tender, and the edges are lightly browned, gently tossing halfway through the cooking time.

Tomatoes Provençal
PREP: 10 MINUTES • COOK TIME: 15 MINUTES • TOTAL: 25 MINUTES • SERVES: 4

Ingredients

4 small ripe tomatoes connected on the vine
¼ teaspoon fine sea salt
¼ teaspoon ground black pepper
½ cup powdered Parmesan cheese (about 1½ ounces)
2 tablespoons chopped fresh parsley
¼ cup minced onions

2 cloves garlic, minced
½ teaspoon chopped fresh thyme leaves
FOR GARNISH:
Fresh parsley leaves
Ground black pepper
Sprig of fresh basil

Instructions

1. Spray the Innsky air fryer basket with avocado oil. Preheat the Innsky air fryer to 350°F. Slice the tops off the tomatoes without removing them from the vine. Do not discard the tops. Use a large spoon to scoop the seeds out of the tomatoes. Sprinkle the insides of the tomatoes with the salt and pepper. In a medium-sized bowl, combine the cheese, parsley, onions, garlic, and thyme. Stir to combine well. Divide the mixture evenly among the tomatoes. Spray avocado oil on the tomatoes and place them in the air fryer basket.

2. Place the tomato tops in the Innsky air fryer basket next to, not on top of, the filled tomatoes. Cook for 15 minutes, or until the filling is golden and the tomatoes are soft yet still holding their shape. Garnish with fresh parsley, ground black pepper, and a sprig of basil. Serve warm, with the tomato tops on the vine. Store leftovers in an airtight container in the refrigerator for up to 4 days. Reheat in a preheated 350°F air fryer for about 3 minutes, until heated through.

Per serving: Calories 68; Fat 3g; Protein 5g; Total carbs 6g; Fiber 1g

Air Fried Rosted Cabbage
PREP: 5 MINUTES • COOK TIME: 7 MINUTES • TOTAL: 12 MINUTES • SERVES: 4

Ingredients

1 head cabbage, sliced in 1-inch-thick ribbons
1 tablespoon olive oil
1 teaspoon salt

1 teaspoon freshly ground black pepper
1 teaspoon garlic powder
1 teaspoon red pepper flakes

Instructions

1. In a large bowl, combine the cabbage, olive oil, salt, pepper, garlic powder, and red pepper flakes. Make sure that the cabbage is thoroughly coated with oil. Place the cabbage in the air fryer basket.

2. Set the temperature of your Innsky AF to 350°F. Set the timer and roast for 4 minutes.
Using tongs, flip the cabbage. Reset the timer and roast for 3 minutes more.

Burrata-Stuffed Tomatoes
PREP: 5 MINUTES • COOK TIME: 5 MINUTES • TOTAL: 10 MINUTES • SERVES: 4

Ingredients

4 medium tomatoes
½ teaspoon fine sea salt
4 (2-ounce) Burrata balls

Fresh basil leaves, for garnish
Extra-virgin olive oil, for drizzling

Instructions

1 Preheat the Innsky air fryer to 300°F.
Core the tomatoes and scoop out the seeds and membranes using a melon baller or spoon. Sprinkle the insides of the tomatoes with the salt. Stuff each tomato with a ball of Burrata.

2 Place in the air fryer and cook for 5 minutes, or until the cheese has softened.
Garnish with basil leaves and drizzle with olive oil. Serve warm.

Per serving: Calories 108; Fat 7g; Protein 6g; Total carbs 5g; Fiber 2g

Spicy Sweet Potato Fries

PREP: 5 MINUTES • COOK TIME: 37 MINUTES • TOTAL: 45 MINUTES • SERVES: 4

Ingredients

2 tbsp. sweet potato fry seasoning mix
2 tbsp. olive oil
2 sweet potatoes

Seasoning Mix:
  2 tbsp. salt

1 tbsp. cayenne pepper
1 tbsp. dried oregano
1 tbsp. fennel
2 tbsp. coriander

Instructions:

1. Slice both ends off sweet potatoes and peel. Slice lengthwise in half and again crosswise to make four pieces from each potato.
   Slice each potato piece into 2-3 slices, then slice into fries.
   Grind together all of seasoning mix ingredients and mix in the salt.
   Ensure the Air fryer is preheated to 350 degrees.
   Toss potato pieces in olive oil, sprinkling with seasoning mix and tossing well to coat thoroughly.
2. Add fries to air fryer basket. Set temperature to 350°F, and set time to 27 minutes. Select START/STOP to begin.
   Take out the basket and turn fries. Turn off air fryer and let cook 10-12 minutes till fries are golden.

Per Serving: Calories: 89; Fat: 14g; Protein: 8gs; Sugar:3g

Creamed Spinach

PREP: 10 MINUTES • COOK TIME: 15 MINUTES • TOTAL: 25 MINUTES • SERVES: 4

Ingredients

Vegetable oil spray
1 (10-ounce) package frozen spinach, thawed and squeezed dry
½ cup chopped onion
2 cloves garlic, minced

4 ounces cream cheese, diced
½ teaspoon ground nutmeg
1 teaspoon kosher salt
1 teaspoon black pepper
½ cup grated Parmesan cheese

Instructions

1. Spray a 6 × 3-inch round heatproof pan with vegetable oil spray.
   In a medium bowl, combine the spinach, onion, garlic, cream cheese, nutmeg, salt, and pepper. Transfer to the prepared pan.
2. Place the pan in the air fryer basket. Set the Innsky air fryer to 350°F for 10 minutes. Open and stir to thoroughly combine the cream cheese and spinach.
   Sprinkle the Parmesan cheese on top. Set the Innsky air fryer to 400°F for 5 minutes, or until the cheese has melted and browned.

Air Fried Buffalo Cauliflower

PREP: 5 MINUTES • COOK TIME: 13 MINUTES • TOTAL: 18 MINUTES • SERVES: 4

Ingredients

4 tablespoons (½ stick) unsalted butter, melted
¼ cup buffalo wing sauce

4 cups cauliflower florets
1 cup panko bread crumbs

Instructions

1 Spray the Innsky air fryer basket with olive oil. In a small bowl, mix the melted butter with the buffalo wing sauce. Put the panko bread crumbs in a separate small bowl.
Dip the cauliflower in the sauce, making sure to coat the top of the cauliflower, then dip the cauliflower in the panko.

2 Place the cauliflower into the greased air fryer basket, being careful not to overcrowd them. Spray the cauliflower generously with olive oil.
Set the temperature of your Innsky AF to 350°F. Set the timer and roast for 7 minutes.
Using tongs, flip the cauliflower. Spray generously with olive oil.
Reset the timer and roast for another 6 minutes.

Per Serving: Calories: 234; Fat: 13g; Saturated fat: 8g; Carbohydrate: 25g; Fiber: 4g; Sugar: 4g; Protein: 4g; Iron: 2mg; Sodium: 333mg

Crispy Brussels Sprouts

PREP: 5 MINUTES • COOK TIME: 8 MINUTES • TOTAL: 13 MINUTES • SERVES: 4

Ingredients

3 tablespoons ghee or coconut oil, melted
1 teaspoon fine sea salt or smoked salt
Dash of lime or lemon juice
Lemon slices, for serving (optional)

Thinly sliced Parmesan cheese, for serving (optional; omit for dairy-free)

Instructions

1. Spray the Innsky air fryer basket with avocado oil. Preheat the Innsky air fryer to 400°F. In a large bowl, toss together the Brussels sprouts, ghee, and salt. Add the lime or lemon juice.

2. Place the Brussels sprouts in the Innsky air fryer basket and cook for 8 minutes, or until crispy, shaking the basket after 5 minutes. Serve with thinly sliced Parmesan and lemon slices, if desired. Best served fresh. Store leftovers in an airtight container in the fridge for up to 5 days.

Broccoli With Parmesan Cheese

PREP: 5 MINUTES • COOK TIME: 4 MINUTES • TOTAL: 9 MINUTES • SERVES: 4

Ingredients

1 pound broccoli florets
2 teaspoons minced garlic

2 tablespoons olive oil
¼ cup grated or shaved Parmesan cheese

Instructions

1. Preheat the Innsky air fryer to 360°F. In a small mixing bowl, mix together the broccoli florets, garlic, olive oil, and Parmesan cheese.

2. Place the broccoli in the Innsky air fryer basket in a single layer and set the timer and steam for 4 minutes.

Creamy Spinach Quiche
PREP: 10 MINUTES • COOK TIME: 20 MINUTES • TOTAL: 30 MINUTES • SERVES: 4

Ingredients

Premade quiche crust, chilled and rolled flat to a
   7-inch round
2 eggs
¼ cup of milk
Pinch of salt and pepper

1 clove of garlic, peeled and finely minced
½ cup of cooked spinach, drained and coarsely
   chopped
¼ cup of shredded mozzarella cheese
¼ cup of shredded cheddar cheese

Instructions

1. Preheat the Innsky air fryer to 360 degrees. Press the premade crust into a 7-inch pie tin, or any appropriately sized glass or ceramic heat-safe dish. Press and trim at the edges if necessary. With a fork, pierce several holes in the dough to allow air circulation and prevent cracking of the crust while cooking. In a mixing bowl, beat the eggs until fluffy and until the yolks and white are evenly combined. Add milk, garlic, spinach, salt and pepper, and half the cheddar and mozzarella cheese to the eggs. Set the rest of the cheese aside for now, and stir the mixture until completely blended. Make sure the spinach is not clumped together, but rather spread among the other ingredients. Pour the mixture into the pie crust, slowly and carefully to avoid splashing. The mixture should almost fill the crust, but not completely – leaving a ¼ inch of crust at the edges.

2. Set the air-fryer timer for 15 minutes. After 15 minutes, the quiche will already be firm and the crust beginning to brown. Sprinkle the rest of the cheddar and mozzarella cheese on top of the quiche filling. ReSet the Innsky air fryer at 360 degrees for 5 minutes.

Mushrooms With Goat Cheese
PREP: 10 MINUTES • COOK TIME: 10 MINUTES • TOTAL: 20 MINUTES • SERVES: 4

Ingredients

3 tablespoons vegetable oil
1 pound mixed mushrooms, trimmed and sliced
1 clove garlic, minced
¼ teaspoon dried thyme

½ teaspoon black pepper
4 ounces goat cheese, diced
3 teaspoons chopped fresh thyme leaves (optional)

Instructions

1. In a 6 × 3-inch round heatproof pan, combine the oil, mushrooms, garlic, dried thyme, and pepper. Stir in the goat cheese.

2. Place the pan in the air fryer basket. Set the Innsky air fryer to 400°F for 10 minutes, stirring halfway through the cooking time. Sprinkle with fresh thyme, if desired.

Caramelized Broccoli
PREP: 5 MINUTES • COOK TIME: 8 MINUTES • TOTAL: 13 MINUTES • SERVES: 4

Ingredients

4 cups broccoli florets
3 tablespoons melted ghee or butter-flavored coconut oil

1½ teaspoons fine sea salt or smoked salt
Mayonnaise, for serving (optional; omit for egg-free)

Instructions

1. Spray the Innsky air fryer basket with avocado oil. Preheat the Innsky air fryer to 400°F. Place the broccoli in a large bowl. Drizzle it with the ghee, toss to coat, and sprinkle it with the salt.

2. Transfer the broccoli to the Innsky air fryer basket and cook for 8 minutes, or until tender and crisp on the edges. Store leftovers in an airtight container in the fridge for up to 4 days or in the freezer for up to a month. Reheat in a preheated 400°F air fryer for 5 minutes, or until crisp.

Sweet Potato French Fries

PREP: 5 MINUTES • COOK TIME: 22 MINUTES • TOTAL: 27 MINUTES • SERVES: 4

Ingredients

2 sweet potatoes

1 teaspoon salt

½ teaspoon freshly ground black pepper

2 teaspoons olive oil.

Instructions

1 Preheat the Innsky air fryer to 380°F. Cut the sweet potatoes lengthwise into ½-inch-thick slices. Then cut each slice into ½-inch-thick fries. In a small mixing bowl, toss the sweet potato fries with the salt, pepper, and olive oil, making sure that all the potatoes are thoroughly coated with oil. Add more oil as needed.

2 Place the potatoes in the air fryer basket. Set the timer and fry for 20 minutes. Shake the basket several times during cooking so that the fries will be evenly cooked and crisp.

Open the air fryer drawer and release the basket. Pour the potatoes into a serving bowl and toss with additional salt and pepper, if desired.

Air Fryer Cauliflower Rice

PREP: 5 MINUTES • COOK TIME: 20 MINUTES • TOTAL: 25 MINUTES • SERVES: 4

Ingredients

Round 1:

tsp. turmeric

1 C. diced carrot

½ C. diced onion

2 tbsp. low-sodium soy sauce

½ block of extra firm tofu

Round 2:

½ C. frozen peas

2 minced garlic cloves

½ C. chopped broccoli

1 tbsp. minced ginger

1 tbsp. rice vinegar

1 ½ tsp. toasted sesame oil

2 tbsp. reduced-sodium soy sauce

3 C. riced cauliflower

Instructions:

1 Crumble tofu in a large bowl and toss with all the Round one ingredients.

2 Preheat the Innsky air fryer to 370 degrees, set temperature to 370°F, and set time to 10 minutes and cook 10 minutes, making sure to shake once. In another bowl, toss ingredients from Round 2 together. Add Round 2 mixture to air fryer and cook another 10 minutes, ensuring to shake 5 minutes in. Enjoy!

Pasta With Mascarpone Mushrooms

PREP: 10 MINUTES • COOK TIME: 15 MINUTES • TOTAL: 25 MINUTES • SERVES: 4

Ingredients

Vegetable oil spray

4 cups sliced mushrooms

1 medium yellow onion, chopped

2 cloves garlic, minced

¼ cup heavy whipping cream or half-and-half

8 ounces mascarpone cheese

1 teaspoon dried thyme

1 teaspoon kosher salt

1 teaspoon black pepper

½ teaspoon red pepper flakes

4 cups cooked konjac noodles, cauliflower rice, linguine, or spaghetti, for serving

½ cup grated Parmesan cheese

Instructions

1 Spray a 7 × 3-inch round heatproof pan with vegetable oil spray. In a medium bowl, combine the mushrooms, onion, garlic, cream, mascarpone, thyme, salt, black pepper, and red pepper flakes. Stir to combine. Transfer the mixture to the prepared pan.

2 Place the pan in the air fryer basket. Set the Innsky air fryer to 350°F for 15 minutes, stirring halfway through the cooking time. Divide the pasta among four shallow bowls. Spoon the mushroom mixture evenly over the pasta. Sprinkle with Parmesan cheese and serve.

Air Fried Hasselback Potatoes

Ingredients

4 russet potatoes
2 tablespoons olive oil
1 teaspoon salt

½ teaspoon freshly ground black pepper
¼ cup grated Parmesan cheese

Instructions

1   Without cutting all the way through the bottom of the potato (so the slices stay connected), cut each potato into ½-inch-wide horizontal slices.
    Brush the potatoes thoroughly with olive oil, being careful to brush in between all the slices. Season with salt and pepper. Place the potatoes in the air fryer basket.

2   Set the temperature of your Innsky AF to 350°F. Set the timer and bake for 20 minutes.
    Brush more olive oil onto the potatoes. Reset the timer and bake for 15 minutes more. Remove the potatoes when they are fork-tender. Sprinkle the cooked potatoes with salt, pepper, and Parmesan cheese.

Marinated Turmeric Cauliflower Steaks

PREP: 5 MINUTES PLUS 20 MINUTES TO MARINATE • COOK TIME: 15 MINUTES • TOTAL: 40 MINUTES • SERVES: 4

Ingredients

¼ cup avocado oil
¼ cup lemon juice
2 cloves garlic, minced
1 teaspoon grated fresh ginger
1 tablespoon turmeric powder

1 teaspoon fine sea salt
1 medium head cauliflower
Full-fat sour cream for serving (optional)
Extra-virgin olive oil, for serving (optional)
Chopped fresh cilantro leaves, for garnish (optional)

Instructions

1   Preheat the Innsky air fryer to 400°F. In a large shallow dish, combine the avocado oil, lemon juice, garlic, ginger, turmeric, and salt. Slice the cauliflower into ½-inch steaks and place them in the marinade. Cover and refrigerate for 20 minutes or overnight. Remove the cauliflower steaks from the marinade and place them in the air fryer basket.

2   Cook for 15 minutes, or until tender and slightly charred on the edges.
    Serve with sour cream and a drizzle of olive oil, and sprinkle with chopped cilantro leaves if desired. Store leftovers in an airtight container in the fridge for up to 4 days or in the freezer for up to a month. Reheat in a preheated 400°F air fryer for 5 minutes, or until warm.

Per serving: Calories 69; Fat 4g; Protein 4g; Total carbs 8g; Fiber 4g

Radishes O'Brien

PREP: 10 MINUTES • COOK TIME: 23 MINUTES • TOTAL: 33 MINUTES • SERVES: 4

Ingredients

2½ cups whole radishes, trimmed, each cut into 8 wedges
1 medium yellow or white onion, diced
1 small green bell pepper, stemmed, seeded, and diced

4 to 6 cloves garlic, thinly sliced
½ to 1 teaspoon kosher salt
½ to 1 teaspoon black pepper
tablespoons coconut oil, melted

Instructions

1   In a large bowl, combine the radishes, onion, bell pepper, garlic, salt, and pepper. Pour the melted oil over the vegetables and mix well to coat.

2

    Scrape the vegetables into the air fryer basket. Set the Innsky air fryer to 350°F for 20 minutes. Increase the temperature to 400°F for 3 minutes to crisp up the edges of the vegetables. Serve hot.

Brown Rice, Spinach and Tofu Frittata
PREP: 5 MINUTES • COOK TIME: 55 MINUTES • TOTAL: 60 MINUTES • SERVES: 4

Ingredients

½ cup baby spinach, chopped
½ cup kale, chopped
½ onion, chopped
½ teaspoon turmeric
1 ¾ cups brown rice, cooked
1 flax egg (1 tablespoon flaxseed meal + 3 tablespoon cold water)
1 package firm tofu
1 tablespoon olive oil
1 yellow pepper, chopped

2 tablespoons soy sauce
2 teaspoons arrowroot powder
2 teaspoons Dijon mustard
2/3 cup almond milk
3 big mushrooms, chopped
3 tablespoons nutritional yeast
4 cloves garlic, crushed
4 spring onions, chopped
a handful of basil leaves, chopped

Instructions:

1. Preheat the Innsky air fryer to 375°F. Grease a pan that will fit inside the air fryer.

    Prepare the frittata crust by mixing the brown rice and flax egg. Press the rice onto the baking dish until you form a crust. Brush with a little oil and cook for 10 minutes.

    Meanwhile, heat olive oil in a skillet over medium flame and sauté the garlic and onions for 2 minutes.

    Add the pepper and mushroom and continue stirring for 3 minutes.

    Stir in the kale, spinach, spring onions, and basil. Remove from the pan and set aside.

    In a food processor, pulse together the tofu, mustard, turmeric, soy sauce, nutritional yeast, vegan milk and arrowroot powder. Pour in a mixing bowl and stir in the sautéed vegetables.

2. Pour the vegan frittata mixture over the rice crust and cook in the air fryer for 40 minutes.

Per Serving: Calories: 226; Fat: 8.05g; Protein: 10.6g

Roasted Rosemary Potatoes
PREP: 5 MINUTES • COOK TIME: 22 MINUTES • TOTAL: 27 MINUTES • SERVES: 4

Ingredients

1½ pounds small red potatoes, cut into 1-inch cubes
2 tablespoons olive oil
1 teaspoon salt

½ teaspoon freshly ground black pepper
1 tablespoon minced garlic
3 tablespoons minced fresh rosemary

Instructions

1   Preheat the Innsky air fryer to 400°F.

    In a medium mixing bowl, combine the diced potatoes, olive oil, salt, pepper, minced garlic, and rosemary and mix well, so the potatoes are thoroughly coated with olive oil.

    Place the potatoes into the Innsky air fryer basket in a single layer.

2   Set the timer and roast for 20 to 22 minutes. Every 5 minutes, remove the air fryer drawer and shake, so the potatoes redistribute in the basket for even cooking.

    Remove the air fryer drawer and release the basket. Pour the potatoes into a large serving bowl, toss with additional salt and pepper, and serve.

Per Serving: Calories: 182; Fat: 7g; Saturated fat: 1g; Carbohydrate: 29g; Fiber: 4g; Sugar: 2g; Protein: 4g; Iron: 2mg; Sodium: 593mg

Caramelized Ranch Cauliflower

PREP: 5 MINUTES • COOK TIME: 12 MINUTES • TOTAL: 17 MINUTES • SERVES: 4

Ingredients

4 cups cauliflower florets
2 tablespoons dried parsley
1 tablespoon plus 1 teaspoon onion powder
2 teaspoons garlic powder
1½ teaspoons dried dill weed

1 teaspoon dried chives
1 teaspoon fine sea salt or smoked salt
1 teaspoon ground black pepper
Ranch Dressing, for serving (optional)

Instructions

1   Preheat the Innsky air fryer to 400°F. Place the cauliflower in a large bowl and spray it with avocado oil. Place the parsley, onion powder, garlic powder, dill weed, chives, salt, and pepper in a small bowl and stir to combine well. Sprinkle the ranch seasoning over the cauliflower.

2   Place the cauliflower in the air fryer and cook for 12 minutes, or until tender and crisp on the edges. Serve with ranch dressing for dipping, if desired. Store leftovers in an airtight container in the fridge for up to 4 days or in the freezer for up to a month. Reheat in a preheated 400°F air fryer for 5 minutes, or until crisp.

Stuffed Mushrooms

PREP: 7 MINUTES • COOK TIME: 8 MINUTES • TOTAL: 15 MINUTES • SERVES: 12

Ingredients

2 Rashers Bacon, Diced
½ Onion, Diced
½ Bell Pepper, Diced
1 Small Carrot, Diced

24 Medium Size Mushrooms
1 cup Shredded Cheddar Plus Extra for the Top
½ cup Sour Cream

Instructions:

1.   Chop the mushrooms stalks finely and fry them up with the bacon, onion, pepper and carrot at 350 ° for 8 minutes. When the veggies are fairly tender, stir in the sour cream & the cheese. Keep on the heat until the cheese has melted and everything is mixed nicely. Now grab the mushroom caps and heap a plop of filling on each one. Place in the fryer basket and top with a little extra cheese.

Roasted Carrots With Harissa Sour Cream

PREP: 10 MINUTES • COOK TIME: 12 MINUTES • TOTAL: 22 MINUTES • SERVES: 4

Ingredients

For the Harissa Sour Cream
½ cup sour cream
1 tablespoon Harissa
For the Carrots
3 cups baby carrots, halved lengthwise
2 tablespoons extra-virgin olive oil

1 teaspoon Ras al Hanout
½ teaspoon kosher salt
1 tablespoon fresh lemon juice
Chopped fresh parsley, for garnish (optional)
Chopped roasted pistachios, for garnish (optional)

Instructions

1   *For the harissa sour cream*: In a small bowl, combine the sour cream and harissa. Whisk until well combined. Cover and chill until ready to serve. *For the carrots*: Place the carrots in a large bowl and drizzle with the olive oil. Toss to coat. Sprinkle with the ras al hanout and salt. Toss again to evenly coat the carrots.

2   Arrange the carrots in the air fryer basket. Set the Innsky air fryer to 400°F for 12 minutes, or until the carrots are tender and lightly charred on the edges, tossing halfway through the cooking time. Transfer the carrots to a serving platter. Drizzle with the lemon juice and harissa sour cream. Sprinkle with the parsley and pistachios, if using.

Air Fried Honey Cornbread

PREP: 5 MINUTES • COOK TIME: 24 MINUTES • TOTAL: 29 MINUTES •SERVES: 4

Ingredients

1 cup all-purpose flour
1 cup yellow cornmeal
½ cup sugar
1 teaspoon salt
2 teaspoons baking powder

1 large egg
1 cup milk
⅓ cup vegetable oil
¼ cup honey

Instructions

1   Spray an air fryer–safe baking pan (square or round) with olive oil or cooking spray. In a large mixing bowl, combine the flour, cornmeal, sugar, salt, baking powder, egg, milk, oil, and honey and mix lightly. Pour the cornbread batter into the prepared pan.

2   Set the temperature of your Innsky AF to 360°F. Set the timer and bake for 20 minutes.
Insert a toothpick into the center of the cornbread to make sure the middle is cooked; if not, bake for another 3 to 4 minutes. Using silicone oven mitts, remove the pan from the air fryer and let cool slightly. Serve warm.

Per Serving: Calories: 594; Fat: 22g; Saturated fat: 5g; Carbohydrate: 94g; Fiber: 3g; Sugar: 46g; Protein: 9g; Iron: 3mg; Sodium: 642mg

Fried Cauliflower Rice

PREP: 5 MINUTES • COOK TIME: 8 MINUTES • TOTAL: 13 MINUTES • SERVES: 4

Ingredients

2 cups cauliflower florets
⅓ cup sliced green onions, plus more for garnish
3 tablespoons wheat-free tamari or coconut aminos
1 clove garlic, smashed to a paste or minced

1 teaspoon grated fresh ginger
1 teaspoon fish sauce or fine sea salt
1 teaspoon lime juice
⅛ teaspoon ground black pepper

Instructions

1   Preheat the Innsky air fryer to 375°F.
Place the cauliflower in a food processor and pulse until it resembles grains of rice.
Place all the ingredients, including the riced cauliflower, in a large bowl and stir well to combine. Transfer the cauliflower mixture to a 6-inch pie pan or a casserole dish that will fit in your air fryer.

2   Cook for 8 minutes, or until soft, shaking halfway through. Garnish with sliced green onions before serving.

Winter Vegetarian Frittata

PREP: 5 MINUTES • COOK TIME: 30 MINUTES • TOTAL: 35 MINUTES • SERVES: 4

Ingredients

1 leek, peeled and thinly sliced into rings
2 cloves garlic, finely minced
3 medium-sized carrots, finely chopped
2 tablespoons olive oil

6 large-sized eggs
Sea salt and ground black pepper, to taste
1/2 teaspoon dried marjoram, finely minced
1/2 cup yellow cheese of choice

Instructions:

1   Sauté the leek, garlic, and carrot in hot olive oil until they are tender and fragrant; reserve. In the meantime, preheat your Air fryer to 330 degrees F.
In a bowl, whisk the eggs along with the salt, ground black pepper, and marjoram.
Then, grease the inside of your baking dish with a nonstick cooking spray. Pour the whisked eggs into the baking dish. Stir in the sautéed carrot mixture. Top with the cheese shreds.

2   Place the baking dish in the Air fryer cooking basket. Cook about 30 minutes and serve warm.

Air Fried Carrots, Yellow Squash & Zucchini
PREP: 5 MINUTES • COOK TIME: 35 MINUTES • TOTAL: 40 MINUTES • SERVES: 4

Ingredients
1 tbsp. chopped tarragon leaves

½ tsp. white pepper

1 tsp. salt

1 pound yellow squash

1 pound zucchini

6 tsp. olive oil

½ pound carrots

Instructions:

1   Stem and root the end of squash and zucchini and cut in ¾-inch half-moons. Peel and cut carrots into 1-inch cubes. Combine carrot cubes with 2 teaspoons of olive oil, tossing to combine.

2   Pour into the Air fryer basket, set temperature to 400°F, and set time to 5 minutes. As carrots cook, drizzle remaining olive oil over squash and zucchini pieces, then season with pepper and salt. Toss well to coat. Add squash and zucchini when the timer for carrots goes off. Cook 30 minutes, making sure to toss 2-3 times during the cooking process. Once done, take out veggies and toss with tarragon. Serve up warm.

Roasted Cauliflower With Cilantro-Jalapeño Sauce
PREP: 15 MINUTES • COOK TIME: 20 MINUTES • TOTAL: 35 MINUTES • SERVES: 4

Ingredients
For the Cauliflower

5 cups cauliflower florets (about 1 large head)

3 tablespoons vegetable oil

½ teaspoon ground cumin

½ teaspoon ground coriander

½ teaspoon kosher salt

For the Sauce

½ cup Greek yogurt or sour cream

¼ cup chopped fresh cilantro

1 jalapeño, seeded and coarsely chopped

4 cloves garlic, peeled

½ teaspoon kosher salt

3 tablespoons water

Instructions

1   For the cauliflower: In a large bowl, combine the cauliflower, oil, cumin, coriander, and salt. Toss to coat.

2   Place the cauliflower in the air fryer basket. Set the Innsky air fryer to 400°F for 20 minutes, stirring halfway through the cooking time. Meanwhile, for the sauce: In a blender, combine the yogurt, cilantro, jalapeño, garlic, and salt. Blend, adding the water as needed to keep the blades moving and to thin the sauce if needed. At the end of cooking time, transfer the cauliflower to a large serving bowl. Pour the sauce over and toss gently to coat. Serve immediately.

Roasted Ratatouille
PREP: 15 MINUTES • COOK TIME: 20 MINUTES • TOTAL: 35 MINUTES • SERVES: 2-3

Ingredients
2 cups ¾-inch cubed peeled eggplant

1 small red, yellow, or orange bell pepper, stemmed, seeded, and diced

1 cup cherry tomatoes

6 to 8 cloves garlic, peeled and halved lengthwise

3 tablespoons olive oil

1 teaspoon dried oregano

½ teaspoon dried thyme

1 teaspoon kosher salt

½ teaspoon black pepper

Instructions

1   In a medium bowl, combine the eggplant, bell pepper, tomatoes, garlic, oil, oregano, thyme, salt, and pepper. Toss to combine.

2   Place the vegetables in the air fryer basket. Set the Innsky air fryer to 400°F for 20 minutes, or until the vegetables are crisp-tender.

Brussels Sprouts with Balsamic Oil

PREP: 5 MINUTES • COOK TIME: 15 MINUTES • TOTAL: 20 MINUTES • SERVES: 4

Ingredients

¼ teaspoon salt

1 tablespoon balsamic vinegar

2 cups Brussels sprouts, halved

3 tablespoons olive oil

Instructions:

1   Preheat the Innsky air fryer for 5 minutes. Mix all ingredients in a bowl until the zucchini fries are well coated.
2   Place in the air fryer basket. Close and cook for 15 minutes for 350°F.

Per Serving: Calories: 82; Fat: 6.8g; Protein: 1.5g

Rosemary & Cheese–Roasted Red Potatoes

PREP: 10 MINUTES • COOK TIME: 15 MINUTES • TOTAL: 25 MINUTES • SERVES: 4

Ingredients

4 cups quartered baby red potatoes

3 tablespoons extra-virgin olive oil

2 teaspoons chopped fresh rosemary

¼ teaspoon garlic powder

Kosher salt and black pepper

¼ cup plus 1 tablespoon finely grated Parmesan cheese

¼ cup chopped fresh parsley

Instructions

1. In a large bowl, toss together the potatoes, olive oil, rosemary, garlic powder, salt and pepper to taste, and ¼ cup of the Parmesan until the potatoes are well coated.

2. Place the seasoned potatoes in the air fryer basket. Set the Innsky air fryer to 400°F for 15 minutes, or until potatoes are tender when pierced with a fork. Transfer the potatoes to a serving platter or bowl. Toss with the remaining 1 tablespoon Parmesan and the parsley.

Air Fried Kale Chips

PREP: 5 MINUTES • COOK TIME: 10 MINUTES • TOTAL: 15 MINUTES • SERVES: 6

Ingredients

¼ tsp. Himalayan salt

3 tbsp. yeast

Avocado oil

1 bunch of kale

Instructions:

1   Rinse kale and with paper towels, dry well.
     Tear kale leaves into large pieces. Remember they will shrink as they cook so good sized pieces are necessary. Place kale pieces in a bowl and spritz with avocado oil till shiny. Sprinkle with salt and yeast. With your hands, toss kale leaves well to combine.
2   Pour half of the kale mixture into the Air fryer, set temperature to 350°F, and set time to 5 minutes. Remove and repeat with another half of kale.

Per Serving: Calories: 55; Fat: 10g; Protein: 1g; Sugar:0g

Spiced Butternut Squash

PREP: 10 MINUTES • COOK TIME: 15 MINUTES • TOTAL: 25 MINUTES • SERVES: 4

Ingredients

4 cups 1-inch-cubed butternut squash

2 tablespoons vegetable oil

1 to 2 tablespoons brown sugar

1 teaspoon Chinese five-spice powder

Instructions

1. In a medium bowl, combine the squash, oil, sugar, and five-spice powder. Toss to coat.
    Place the squash in the air fryer basket.
2. Set the Innsky air fryer to 400°F for 15 minutes or until tender.

Garlic Thyme Mushrooms

PREP: 5 MINUTES • COOK TIME: 10 MINUTES • TOTAL: 15 MINUTES • SERVES: 4

Ingredients

3 tablespoons unsalted butter (or butter-flavored coconut oil for dairy-free), melted

1 (8-ounce) package button mushrooms, sliced

2 cloves garlic, minced

3 sprigs fresh thyme leaves, plus more for garnish

½ teaspoon fine sea salt

Instructions

1. Spray the Innsky air fryer basket with avocado oil. Preheat the Innsky air fryer to 400°F.
   Place all the ingredients in a medium-sized bowl. Use a spoon or your hands to coat the mushroom slices.
2. Place the mushrooms in the Innsky air fryer basket in one layer; work in batches if necessary. Cook for 10 minutes, or until slightly crispy and brown. Garnish with thyme sprigs before serving.
   Store leftovers in an airtight container in the fridge for up to 5 days or in the freezer for up to a month. Reheat in a preheated 350°F air fryer for 5 minutes, or until heated through.

Per serving: Calories 82; Fat 9g; Protein 1g; Total carbs 1g; Fiber 0.2g

Zucchini Omelet

PREP: 10 MINUTES • COOK TIME: 10 MINUTES • TOTAL: 20 MINUTES • SERVES: 2

Ingredients

1 teaspoon butter

1 zucchini, julienned

4 eggs

¼ teaspoon fresh basil, chopped

¼ teaspoon red pepper flakes, crushed

Salt and freshly ground black pepper, to taste

Instructions:

1. Preheat the Innsky air fryer to 355 degrees F. In a skillet, melt butter on medium heat. Add zucchini and cook for about 3-4 minutes. In a bowl, add the eggs, basil, red pepper flakes, salt and black pepper and beat well. Add cooked zucchini and gently, stir to combine.
2. Transfer the mixture into the Air fryer pan. Cook for about 10 minutes or till done completely.

Cheesy Cauliflower Fritters

PREP: 10 MINUTES • COOK TIME: 7 MINUTES • TOTAL: 17 MINUTES • SERVES: 8

Ingredients

½ C. chopped parsley

1 C. Italian breadcrumbs

1/3 C. shredded mozzarella cheese

1/3 C. shredded sharp cheddar cheese

1 egg

2 minced garlic cloves

3 chopped scallions

1 head of cauliflower

Instructions:

1. Cut cauliflower up into florets. Wash well and pat dry. Place into a food processor and pulse 20-30 seconds till it looks like rice.
   Place cauliflower rice in a bowl and mix with pepper, salt, egg, cheeses, breadcrumbs, garlic, and scallions. With hands, form 15 patties of the mixture. Add more breadcrumbs if needed.
2. With olive oil, spritz patties, and place into your Air fryer in a single layer. Set temperature to 390°F, and set time to 7 minutes, flipping after 7 minutes.

Per Serving: Calories: 209; Fat: 17g; Protein: 6g; Sugar:0.5

Spiced Glazed Carrots
PREP: 10 MINUTES • COOK TIME: 30 MINUTES • TOTAL: 40 MINUTES • SERVES: 4

Ingredients
Vegetable oil spray
4 cups frozen sliced carrots (do not thaw)
2 tablespoons brown sugar
2 tablespoons water
½ teaspoon ground cumin

½ teaspoon ground cinnamon
¼ teaspoon kosher salt
2 tablespoons coconut oil
Chopped fresh parsley, for garnish

Instructions
1. Spray a 6 × 4-inch round heatproof pan with vegetable oil spray.
   In a medium bowl, combine the carrots, brown sugar, water, cumin, cinnamon, and salt. Toss to coat. Transfer to the prepared pan. Dot the carrots with the coconut oil, distributing it evenly across the pan. Cover the pan with foil.
2. Place the pan in the air fryer basket. Set the Innsky air fryer to 400°F for 10 minutes. Remove the foil and stir well. Place the uncovered pan back in the air fryer. Set the Innsky air fryer to 400°F for 2 minutes, or until the glaze is bubbling and the carrots are cooked through. Garnish with parsley and serve.

Spinach & Cheese–Stuffed Tomatoes
PREP: 20 MINUTES • COOK TIME: 15 MINUTES • TOTAL: 35 MINUTES • SERVES: 2

Ingredients
4 ripe beefsteak tomatoes
¾ teaspoon black pepper
½ teaspoon kosher salt
1 (10-ounce) package frozen chopped spinach, thawed and squeezed dry

1 (5.2-ounce) package garlic-and-herb Boursin cheese
3 tablespoons sour cream
½ cup finely grated Parmesan cheese

Instructions
1. Cut the tops off the tomatoes. Using a small spoon, carefully remove and discard the pulp. Season the insides with ½ teaspoon of the black pepper and ¼ teaspoon of the salt. Invert the tomatoes onto paper towels and allow to drain while you make the filling.
   Meanwhile, in a medium bowl, combine the spinach, Boursin cheese, sour cream, ¼ cup of the Parmesan, and the remaining ¼ teaspoon salt and ¼ teaspoon pepper. Stir until ingredients are well combined. Divide the filling among the tomatoes. Top with the remaining ¼ cup Parmesan.
2. Place the tomatoes in the air fryer basket. Set the Innsky air fryer to 350°F for 15 minutes, or until the filling is hot.

Spinach Artichoke Tart

PREP: 10 MINUTES • COOK TIME: 40 MINUTES • TOTAL: 50 MINUTES • SERVES: 6

Ingredients

CRUST:
1 cup blanched almond flour
1 cup grated Parmesan cheese
1 large egg
FILLING:
4 ounces cream cheese (½ cup), softened
1 (8-ounce) package frozen chopped spinach, thawed and drained

½ cup artichoke hearts, drained and chopped
⅓ cup shredded Parmesan cheese, plus more for topping
1 large egg
1 clove garlic, minced
¼ teaspoon fine sea salt

Instructions

1. Preheat the Innsky air fryer to 350°F.
   Make the crust: Place the almond flour and cheese in a large bowl and mix until well combined. Add the egg and mix until the dough is well combined and stiff.
   Press the dough into a 6-inch pie pan. Bake for 8 to 10 minutes, until it starts to brown lightly.
   Make the filling: Place the cream cheese in a large bowl and stir to break it up. Add the spinach, artichoke hearts, cheese, egg, garlic, and salt. Stir well to combine.
   Pour the spinach mixture into the prebaked crust and sprinkle with additional Parmesan.
2. Place in the air fryer and cook for 25 to 30 minutes, until cooked through.
   Store leftovers in an airtight container in the fridge for up to 4 days or in the freezer for up to a month. Reheat in a preheated 350°F air fryer for 5 minutes, or until heated through.

Per serving: Calories 228; Fat 7g; Protein 14g; Total carbs 6g; Fiber 2g

Cauliflower Bites

PREP: 10 MINUTES • COOK TIME: 18 MINUTES • TOTAL: 28 MINUTES • SERVES: 4

Ingredients

1 Head Cauliflower, cut into small florets
Tsps Garlic Powder
Pinch of Salt and Pepper

1 Tbsp Butter, melted
1/2 Cup Chili Sauce
Olive Oil

Instructions:

1. Place cauliflower into a bowl and pour oil over florets to lightly cover.
   Season florets with salt, pepper and the garlic powder and toss well.
2. Place florets into the Air fryer at 350 degrees for 14 minutes. Remove cauliflower from the Air Fryer. Combine the melted butter with the chili sauce. Pour over the florets so that they are well coated. Return to the Air fryer and cook for additional 3 to 4 minutes. Serve as a side or with ranch or cheese dip as a snack

Russet & Sweet Potato Gratin
PREP: 15 MINUTES • COOK TIME: 45 MINUTES • TOTAL: 60 MINUTES • SERVES: 4

Ingredients

Vegetable oil spray
1 cup heavy whipping cream
¼ cup roughly chopped onion
3 cloves garlic, peeled
1 teaspoon chopped fresh thyme leaves
½ teaspoon kosher salt
½ teaspoon black pepper

1 medium russet potato, peeled and very thinly sliced (about 1½ cups)
1 medium sweet potato, peeled and very thinly sliced (about 1½ cups)
1 tablespoon vegetable oil
¼ cup grated Parmesan cheese

Instructions

1. Spray a 6 × 3-inch round heatproof pan with vegetable oil spray; set aside.
   In a blender, combine the cream, onion, garlic, thyme, salt, and pepper. Blend until smooth. In a large bowl, combine the sliced russet and sweet potatoes. Drizzle with the oil and toss to coat. Transfer the potatoes to the prepared pan. Pour the cream mixture over the top of the potatoes.
2. Cover the pan with foil and place in the air fryer basket. Set the Innsky air fryer to 350°F for 40 minutes, or until potatoes are nearly tender. Uncover and sprinkle with Parmesan cheese. Set the Innsky air fryer to 400°F for 5 minutes, or until the potatoes are bubbly with a golden-brown crust.

Buttered Carrot-Zucchini with Mayo
PREP: 10 MINUTES • COOK TIME: 25 MINUTES • TOTAL: 35 MINUTES • SERVES: 4

Ingredients

1 tablespoon grated onion
2 tablespoons butter, melted
1/2-pound carrots, sliced
1-1/2 zucchinis, sliced
1/4 cup water

1/4 cup mayonnaise
1/4 teaspoon prepared horseradish
1/4 teaspoon salt
1/4 teaspoon ground black pepper
1/4 cup Italian bread crumbs

Instructions:

1. Lightly grease baking pan of air fryer with cooking spray. Add carrots. For 8 minutes, cook on 360°F. Add zucchini and continue cooking for another 5 minutes.
   Meanwhile, in a bowl whisk well pepper, salt, horseradish, onion, mayonnaise, and water. Pour into pan of veggies. Toss well to coat.
   In a small bowl mix melted butter and bread crumbs. Sprinkle over veggies.
2. Cook for 10 minutes at 390°F, until tops are lightly browned.
   Serve and enjoy.

Sweet Potato Fries With Aji Criollo Mayo
PREP: 10 MINUTES • COOK TIME: 20 MINUTES • TOTAL: 30 MINUTES • SERVES: 2

Ingredients

For the Fries

1 large sweet potato, peeled

2 tablespoons extra-virgin olive oil

½ teaspoon smoked paprika

½ teaspoon garlic powder

½ teaspoon onion powder

¼ teaspoon kosher salt

¼ teaspoon black pepper

For the Aji Criollo Mayonnaise

½ cup mayonnaise

2 tablespoons fresh lime juice

2 teaspoons cider vinegar

½ bunch fresh cilantro, roughly chopped

¼ bunch fresh parsley, roughly chopped

1 green onion (white and green parts), chopped

2 jalapeños, seeded and chopped

2 cloves garlic, minced

¼ teaspoon kosher salt

Instructions

1. For the fries: Cut the potato lengthwise into ¼-inch-thick slices. Lay each slice flat and cut lengthwise into fries about ¼ inch thick. In a medium bowl, toss together the potatoes, olive oil, paprika, garlic powder, onion powder, salt, and pepper until well coated.

2. Place the fries in a single layer in the air fryer basket. (If they won't fit in a single layer, set a rack or trivet on top of the bottom layer of potatoes and place the rest of the potatoes on the rack, or cook in multiple batches.) Set the Innsky air fryer to 400°F for 20 minutes, shaking halfway through the cooking time, or until the fries are tender and lightly browned. Meanwhile, for the mayonnaise: In a blender, combine the mayonnaise, lime juice, vinegar, cilantro, parsley, green onion, jalapeños, garlic, and salt. Blend until smooth. Turn the fries out onto a serving platter.

Avocado Fries
PREP: 10 MINUTES • COOK TIME: 7 MINUTES • TOTAL: 17 MINUTES • SERVES: 6

Ingredients

1 avocado

½ tsp. salt

½ C. panko breadcrumbs

Bean liquid (aquafaba) from a 15-ounce can of white or garbanzo beans

Instructions:

1. Peel, pit, and slice up avocado.
Toss salt and breadcrumbs together in a bowl. Place aquafaba into another bowl.
Dredge slices of avocado first in aquafaba and then in panko, making sure you get an even coating.

2. Place coated avocado slices into a single layer in the Air fryer. Set temperature to 390°F, and set time to 5 minutes. Serve with your favorite keto dipping sauce.

Sweet & Crispy Roasted Pearl Onions
PREP: 5 MINUTES • COOK TIME: 18 MINUTES • TOTAL: 23 MINUTES • SERVES: 3

Ingredients

1 (14.5-ounce) package frozen pearl onions (do not thaw)

2 tablespoons extra-virgin olive oil

2 tablespoons balsamic vinegar

2 teaspoons finely chopped fresh rosemary

½ teaspoon kosher salt

¼ teaspoon black pepper

Instructions

1. In a medium bowl, combine the onions, olive oil, vinegar, rosemary, salt, and pepper until well coated.

2. Transfer the onions to the air fryer basket. Set the Innsky air fryer to 400°F for 18 minutes, or until the onions are tender and lightly charred, stirring once or twice during the cooking time.

Zucchini & Tomato Salad
PREP: 10 MINUTES • COOK TIME: 10 MINUTES • TOTAL: 20 MINUTES • SERVES: 2-3

Ingredients

For the Salad
3 tablespoons olive oil
2 teaspoons Lebanese Seven-Spice Mix
½ teaspoon ground cumin (optional)
1 medium zucchini, halved lengthwise and cut into chunks (about 2 cups)
1 cup cherry tomatoes
Olive oil spray

For the Dressing
¼ cup olive oil
2 tablespoons fresh lemon juice
¼ cup chopped fresh parsley
2 tablespoons chopped fresh mint
½ to 1 teaspoon Lebanese Seven-Spice Mix
½ teaspoon kosher salt

Instructions

1. For the salad: In a medium bowl, whisk together the oil, spice mix, and cumin if using. Add the zucchini and tomatoes and toss to combine.
2. Spray the Innsky air fryer basket with olive oil spray. Place the zucchini and tomatoes in the air fryer basket. Set the Innsky air fryer to 400°F for 10 minutes. Transfer the vegetables to a serving bowl and let cool. Meanwhile, for the dressing: In a small jar with a lid, combine all the ingredients and shake vigorously to combine. Pour the dressing over the cooled vegetables, toss until well coated, and serve.

Roasted Vegetables Salad
PREP: 5 MINUTES • COOK TIME: 85 MINUTES • TOTAL: 90 MINUTES • SERVES: 5

Ingredients

3 eggplants
1 tbsp of olive oil
3 medium zucchini
1 tbsp of olive oil
4 large tomatoes, cut them in eighths
4 cups of one shaped pasta

2 peppers of any color
1 cup of sliced tomatoes cut into small cubes
2 teaspoon of salt substitute
8 tbsp of grated parmesan cheese
½ cup of Italian dressing
Leaves of fresh basil

Instructions:

1. Wash your eggplant and slice it off then discard the green end. Make sure not to peel. Slice your eggplant into1/2 inch of thick rounds. 1/2 inch)
Pour 1tbsp of olive oil on the eggplant round.
2. Put the eggplants in the basket of the Air fryer and then toss it in the air fryer. Cook the eggplants for 40 minutes. Set the heat to 360 ° F
Meanwhile, wash your zucchini and slice it then discard the green end. But do not peel it.
Slice the Zucchini into thick rounds of ½ inch each.
In the basket of the Air Fryer, toss your ingredients and add 1 tbsp of olive oil.
3. Cook the zucchini for 25 minutes on a heat of 360° F and when the time is off set it aside.
Wash and cut the tomatoes.
4. Arrange your tomatoes in the basket of the Air fryer. Set the timer to 30 minutes. Set the heat to 350° F
When the time is off, cook your pasta according to the pasta guiding Instructions, empty it into a colander. Run the cold water on it and wash it and drain the pasta and put it aside.
Meanwhile, wash and chop your peppers and place it in a bow. Wash and thinly slice your cherry tomatoes and add it to the bowl. Add your roasted veggies.
Add the pasta, a pinch of salt, the topping dressing, add the basil and the parm and toss everything together. (It is better to mix with your hands). Set the ingredients together in the refrigerator, and let it chill. Serve your salad and enjoy it.

Cheddar, Squash 'n Zucchini Casserole
PREP: 5 MINUTES • COOK TIME: 30 MINUTES • TOTAL: 35 MINUTES • SERVES: 4

Ingredients
1 egg
5 saltine crackers, or as needed, crushed
2 tablespoons bread crumbs
1/2-pound yellow squash, sliced
1/2-pound zucchini, sliced
1/2 cup shredded Cheddar cheese

1-1/2 teaspoons white sugar
1/2 teaspoon salt
1/4 onion, diced
1/4 cup biscuit baking mix
1/4 cup butter

Instructions:
1   Lightly grease baking pan of air fryer with cooking spray. Add onion, zucchini, and yellow squash. Cover pan with foil and for 15 minutes, cook on 360° F or until tender.

Stir in salt, sugar, egg, butter, baking mix, and cheddar cheese. Mix well. Fold in crushed crackers. Top with bread crumbs.

2   Cook for 15 minutes at 390° F until tops are lightly browned.

Serve and enjoy.

Zucchini Parmesan Chips
PREP: 10 MINUTES • COOK TIME: 8 MINUTES • TOTAL: 18 MINUTES • SERVES: 10

Ingredients
½ tsp. paprika
½ C. grated parmesan cheese
½ C. Italian breadcrumbs

1 lightly beaten egg
2 thinly sliced zucchinis

Instructions:
1   Use a very sharp knife or mandolin slicer to slice zucchini as thinly as you can. Pat off extra moisture. Beat egg with a pinch of pepper and salt and a bit of water.

Combine paprika, cheese, and breadcrumbs in a bowl. Dip slices of zucchini into the egg mixture and then into breadcrumb mixture. Press gently to coat.

2   With olive oil cooking spray, mist coated zucchini slices. Place into your Air fryer in a single layer. Set temperature to 350°F, and set time to 8 minutes. Sprinkle with salt and serve with salsa.

Crispy Roasted Broccoli
PREP: 10 MINUTES • COOK TIME: 8 MINUTES • TOTAL: 18 MINUTES • SERVES: 2

Ingredients
¼ tsp. Masala
½ tsp. red chili powder
½ tsp. salt
¼ tsp. turmeric powder

1 tbsp. chickpea flour
2 tbsp. yogurt
1 pound broccoli

Instructions:
1   Cut broccoli up into florets. Soak in a bowl of water with 2 teaspoons of salt for at least half an hour to remove impurities. Take out broccoli florets from water and let drain. Wipe down thoroughly. Mix all other ingredients together to create a marinade.

Toss broccoli florets in the marinade. Cover and chill 15-30 minutes.

2   Preheat the Innsky air fryer to 390 degrees. Place marinated broccoli florets into the fryer, set temperature to 350°F, and set time to 10 minutes. Florets will be crispy when done.

Jalapeño Cheese Balls
PREP: 10 MINUTES • COOK TIME: 8 MINUTES • TOTAL: 18 MINUTES • SERVES: 12

Ingredients

4 ounces cream cheese
⅓ cup shredded mozzarella cheese
⅓ cup shredded Cheddar cheese
2 jalapeños, finely chopped
½ cup bread crumbs

2 eggs
½ cup all-purpose flour
Salt
Pepper
Cooking oil

Instructions:

1   In a medium bowl, combine the cream cheese, mozzarella, Cheddar, and jalapeños. Mix well. Form the cheese mixture into balls about an inch thick. Using a small ice cream scoop works well. Arrange the cheese balls on a sheet pan and place in the freezer for 15 minutes. This will help the cheese balls maintain their shape while frying. Spray the Innsky air fryer basket with cooking oil. Place the bread crumbs in a small bowl. In another small bowl, beat the eggs. In a third small bowl, combine the flour with salt and pepper to taste, and mix well. Remove the cheese balls from the freezer. Dip the cheese balls in the flour, then the eggs, and then the bread crumbs.

2   Place the cheese balls in the Air fryer. Spray with cooking oil. Cook for 8 minutes. Open the air fryer and flip the cheese balls. I recommend flipping them instead of shaking so the balls maintain their form. Cook an additional 4 minutes. Cool before serving.

Creamy And Cheese Broccoli Bake
PREP: 5 MINUTES • COOK TIME: 30 MINUTES • TOTAL: 35 MINUTES • SERVES: 2

Ingredients

1-pound fresh broccoli, coarsely chopped
2 tablespoons all-purpose flour
salt to taste
1 tablespoon dry bread crumbs, or to taste
1/2 large onion, coarsely chopped

1/2 (14 ounce) can evaporated milk, divided
1/2 cup cubed sharp Cheddar cheese
1-1/2 teaspoons butter, or to taste
1/4 cup water

Instructions:

1   Lightly grease baking pan of air fryer with cooking spray. Mix in half of the milk and flour in pan and for 5 minutes, cook on 360°F. Halfway through cooking time, mix well. Add broccoli and remaining milk. Mix well and cook for another 5 minutes. Stir in cheese and mix well until melted. In a small bowl mix well, butter and bread crumbs. Sprinkle on top of broccoli.

2   Cook for 20 minutes at 360°F until tops are lightly browned. Serve and enjoy.

Buffalo Cauliflower
PREP: 5 MINUTES • COOK TIME: 15 MINUTES • TOTAL: 20 MINUTES • SERVES: 2

Ingredients

Cauliflower:
1 C. panko breadcrumbs
1 tsp. salt
4 C. cauliflower florets

Buffalo Coating:
¼ C. Vegan Buffalo sauce
¼ C. melted vegan butter

Instructions:

1   Melt butter in microwave and whisk in buffalo sauce.
Dip each cauliflower floret into buffalo mixture, ensuring it gets coated well. Hold over a bowl till floret is done dripping.
Mix breadcrumbs with salt.

2   Dredge dipped florets into breadcrumbs and place into air fryer. Set temperature to 350°F, and set time to 15 minutes. When slightly browned, they are ready to eat. Serve with your favorite keto dipping sauce.

Coconut Battered Cauliflower Bites
PREP: 5 MINUTES • COOK TIME: 20 MINUTES • TOTAL: 25 MINUTES • SERVES: 4

Ingredients
salt and pepper to taste
1 flax egg (1 tablespoon flaxseed meal + 3 tablespoon water)
1 small cauliflower, cut into florets
1 teaspoon mixed spice
½ teaspoon mustard powder

2 tablespoons maple syrup
1 clove of garlic, minced
2 tablespoons soy sauce
1/3 cup oats flour
1/3 cup plain flour
1/3 cup desiccated coconut

Instructions:
1   Preheat the Innsky air fryer to 400°F. In a mixing bowl, mix together oats, flour, and desiccated coconut. Season with salt and pepper to taste. Set aside. In another bowl, place the flax egg and add a pinch of salt to taste. Set aside. Season the cauliflower with mixed spice and mustard powder. Dredge the florets in the flax egg first then in the flour mixture.

2   Place inside the air fryer and cook for 15 minutes. Meanwhile, place the maple syrup, garlic, and soy sauce in a sauce pan and heat over medium flame. Bring to a boil and adjust the heat to low until the sauce thickens. After 15 minutes, take out the florets from the air fryer and place them in the saucepan. Toss to coat the florets and place inside the air fryer and cook for another 5 minutes.

Crispy Jalapeno Coins
PREP: 10 MINUTES • COOK TIME: 5 MINUTES • TOTAL: 15 MINUTES • SERVES: 2

Ingredients
1 egg
2-3 tbsp. coconut flour
1 sliced and seeded jalapeno
Pinch of garlic powder

Pinch of onion powder
Pinch of Cajun seasoning (optional)
Pinch of pepper and salt

Instructions:
1   Ensure your air fryer is preheated to 400 degrees.
    Mix together all dry ingredients. Pat jalapeno slices dry. Dip coins into egg wash and then into dry mixture. Toss to thoroughly coat. Add coated jalapeno slices to air fryer in a singular layer. Spray with olive oil.
2   Set temperature to 350°F, and set time to 5 minutes. Cook just till crispy.

Vegetable Egg Rolls
PREP: 15 MINUTES • COOK TIME: 10 MINUTES • TOTAL: 25 MINUTES • SERVES: 8

Ingredients
½ cup chopped mushrooms
½ cup grated carrots
½ cup chopped zucchini
2 green onions, chopped

2 tablespoons low-sodium soy sauce
8 egg roll wrappers
1 tablespoon cornstarch
1 egg, beaten

Instructions:
1   In a medium bowl, combine the mushrooms, carrots, zucchini, green onions, and soy sauce, and stir together. Place the egg roll wrappers on a work surface. Top each with about 3 tablespoons of the vegetable mixture. In a small bowl, combine the cornstarch and egg and mix well. Brush some of this mixture on the edges of the egg roll wrappers. Roll up the wrappers, enclosing the vegetable filling. Brush some of the egg mixture on the outside of the egg rolls to seal.
2   Air-fry for 7 to 10 minutes or until the egg rolls are brown and crunchy.

Per Serving: Calories: 112; Fat: 1g; Protein:4g; Fiber:1g

Veggies on Toast

PREP: 12 MINUTES • COOK TIME: 11 MINUTES • TOTAL: 23 MINUTES • SERVES: 4

Ingredients

1 red bell pepper, cut into ½-inch strips
1 cup sliced button or cremini mushrooms
1 small yellow squash, sliced
2 green onions, cut into ½-inch slices

Extra light olive oil for misting
4 to 6 pieces sliced French or Italian bread
2 tablespoons softened butter
½ cup soft goat cheese

Instructions:

1. Combine the red pepper, mushrooms, squash, and green onions in the Air fryer and mist with oil.
2. Roast for 7 to 9 minutes or until the vegetables are tender, shaking the basket once during cooking time. Remove the vegetables from the basket and set aside. Spread the bread with butter and place in the Air fryer, butter-side up.
3. Toast for 2 to 4 minutes or until golden brown. Spread the goat cheese on the toasted bread and top with the vegetables; serve warm.

Per Serving: Calories: 162; Fat: 11g; Protein:7g; Fiber:2g

Crisped Baked Cheese Stuffed Chile Pepper

PREP: 10 MINUTES • COOK TIME: 30 MINUTES • TOTAL: 40 MINUTES • SERVES: 3

Ingredients

1 (7 ounce) can whole green Chile peppers, drained
1 egg, beaten
1 tablespoon all-purpose flour
1/2 (5 ounce) can evaporated milk

1/2 (8 ounce) can tomato sauce
1/4-pound Monterey Jack cheese, shredded
1/4-pound Longhorn or Cheddar cheese, shredded
1/4 cup milk

Instructions:

1   Lightly grease baking pan of air fryer with cooking spray. Evenly spread chilies and sprinkle cheddar and Jack cheese on top.

In a bowl whisk well flour, milk, and eggs. Pour over chilies.

2   For 20 minutes, cook on 360°F. Add tomato sauce on top. Cook for 10 minutes at 390°F until tops are lightly browned. Serve and enjoy.

Per Serving: Calories: 392; Fat: 27.6g; Protein:23.9g

Jicama Fries

PREP: 10 MINUTES • COOK TIME: 5 MINUTES • TOTAL: 15 MINUTES • SERVES: 8

Ingredients

1 tbsp. dried thyme
¾ C. arrowroot flour

½ large Jicama
Eggs

Instructions:

1   Sliced jicama into fries.

Whisk eggs together and pour over fries. Toss to coat.

Mix a pinch of salt, thyme, and arrowroot flour together. Toss egg-coated jicama into dry mixture, tossing to coat well.

2   Spray the Innsky air fryer basket with olive oil and add fries. Set temperature to 350°F, and set time to 5 minutes. Toss halfway into the cooking process.

Per Serving: Calories: 211; Fat: 19g; Protein:9g

Jumbo Stuffed Mushrooms

PREP: 10 MINUTES • COOK TIME: 20 MINUTES • TOTAL: 30 MINUTES • SERVES: 4

Ingredients

4 jumbo portobello mushrooms

1 tablespoon olive oil

¼ cup ricotta cheese

5 tablespoons Parmesan cheese, divided

1 cup frozen chopped spinach, thawed and drained

⅓ cup bread crumbs

¼ teaspoon minced fresh rosemary

Instructions:

1. Wipe the mushrooms with a damp cloth. Remove the stems and discard. Using a spoon, gently scrape out most of the gills.

   Rub the mushrooms with the olive oil.

2. Put in the Air fryer basket, hollow side up, and bake for 3 minutes. Carefully remove the mushroom caps, because they will contain liquid. Drain the liquid out of the caps. In a medium bowl, combine the ricotta, 3 tablespoons of Parmesan cheese, spinach, bread crumbs, and rosemary, and mix well. Stuff this mixture into the drained mushroom caps. Sprinkle with the remaining 2 tablespoons of Parmesan cheese. Put the mushroom caps back into the basket and bake for 4 to 6 minutes or until the filling is hot and the mushroom caps are tender.

Per Serving: Calories: 117; Fat: 7g; Protein:7g; Fiber:1g

Air Fryer Brussels Sprouts

PREP: 10 MINUTES • COOK TIME: 10 MINUTES • TOTAL: 20 MINUTES • SERVES: 8

Ingredients

¼ tsp. salt

1 tbsp. balsamic vinegar

1 tbsp. olive oil

2 C. Brussels sprouts

Instructions:

1. Cut Brussels sprouts in half lengthwise. Toss with salt, vinegar, and olive oil till coated thoroughly.

2. Add coated sprouts to the Air fryer, close crisping lid, set temperature to 400°F, and set time to 10 minutes. Shake after 5 minutes of cooking.

   Brussels sprouts are ready to devour when brown and crisp!

Per Serving: Calories: 118; Fat: 9g; Protein:11g

Spaghetti Squash Tots

PREP: 10 MINUTES • COOK TIME: 15 MINUTES • TOTAL: 25 MINUTES • SERVES: 8

Ingredients

¼ tsp. pepper

½ tsp. salt

1 thinly sliced scallion

1 spaghetti squash

Instructions:

1. Wash and cut the squash in half lengthwise. Scrape out the seeds.

   With a fork, remove spaghetti meat by strands and throw out skins.

   In a clean towel, toss in squash and wring out as much moisture as possible. Place in a bowl and with a knife slice through meat a few times to cut up smaller.

   Add pepper, salt, and scallions to squash and mix well.

2. Create "tot" shapes with your hands and place in the Air fryer. Spray with olive oil. Set temperature to 350°F, and set time to 15 minutes. Cook until golden and crispy!

Per Serving: Calories: 231; Fat: 18g; Protein:5g; Sugar:0g

Crispy And Healthy Avocado Fingers

PREP: 10 MINUTES • COOK TIME: 10 MINUTES • TOTAL: 20 MINUTES • SERVES: 4

Ingredients

½ cup panko breadcrumbs
½ teaspoon salt

1 pitted Haas avocado, peeled and sliced
liquid from 1 can white beans or aquafaba

Instructions:

1 Preheat the Innsky air fryer at 350°F. In a shallow bowl, toss the breadcrumbs and salt until well combined. Dredge the avocado slices first with the aquafaba then in the breadcrumb mixture. Place the avocado slices in a single layer inside the air fryer basket.
2 Cook for 10 minutes and shake halfway through the cooking time.

Per Serving: Calories: 51; Fat: 7.5g; Protein:1.39g

Onion Rings

PREP: 10 MINUTES • COOK TIME: 10 MINUTES • TOTAL: 20 MINUTES • SERVES: 4

Ingredients

1 large spanish onion
1/2 cup buttermilk
2 eggs, lightly beaten
3/4 cups unbleached all-purpose flour

3/4 cups panko bread crumbs
1/2 teaspoon baking powder
1/2 teaspoon Cayenne pepper, to taste
Salt

Instructions:

1 Start by cutting your onion into 1/2 thick rings and separate. Smaller pieces can be discarded or saved for other recipes.
Beat the eggs in a large bowl and mix in the buttermilk, then set it aside.
In another bowl combine flour, pepper, bread crumbs, and baking powder.
Use a large spoon to dip a whole ring in the buttermilk, then pull it through the flour mix on both sides to completely coat the ring.
2 Cook about 8 rings at a time in your Air fryer for 8-10 minutes at 360 degrees shaking half way through.

Per Serving: Calories: 225; Fat: 3.8g; Protein:19g; Fiber:2.4g

Cinnamon Butternut Squash Fries

PREP: 5 MINUTES • COOK TIME: 10 MINUTES • TOTAL: 15 MINUTES • SERVES: 8

Ingredients

1 pinch of salt
1 tbsp. powdered unprocessed sugar
½ tsp. nutmeg

2 tsp. cinnamon
1 tbsp. coconut oil
10 ounces pre-cut butternut squash fries

Instructions:

1 In a plastic bag, pour in all ingredients. Coat fries with other components till coated and sugar is dissolved.
2 Spread coated fries into a single layer in the Air fryer. Set temperature to 390°F, and set time to 10 minutes. Cook until crispy.

Per Serving: Calories: 175; Fat: 8g; Protein:1g

Poultry Recipes

Korean Chicken Wings
PREP: 5 MINUTES • COOK TIME: 10 MINUTES • TOTAL: 15 MINUTES • SERVES: 8

Ingredients
Wings:
1 tsp. pepper
1 tsp. salt
2 pounds chicken wings
Sauce:
2 packets Splenda
1 tbsp. minced garlic
1 tbsp. minced ginger

1 tbsp. sesame oil
1 tsp. agave nectar
1 tbsp. mayo
2 tbsp. gochujang
Finishing:
¼ C. chopped green onions
2 tsp. sesame seeds

Instructions:
1 Ensure air fryer is preheated to 400 degrees.
   Line a small pan with foil and place a rack onto the pan, then place into air fryer.
   Season wings with pepper and salt and place onto the rack.
2   Set temperature to 160°F, and set time to 20 minutes and air fry 20 minutes, turning at 10 minutes. As chicken air fries, mix together all the sauce components.
   Once a thermometer says that the chicken has reached 160 degrees, take out wings and place into a bowl. Pour half of the sauce mixture over wings, tossing well to coat. Put coated wings back into air fryer for 5 minutes or till they reach 165 degrees. Remove and sprinkle with green onions and sesame seeds. Dip into extra sauce.

Per Serving: Calories: 356; Fat: 26g; Protein:23g; Sugar:2g

Buffalo Chicken Wings
PREP: 10 MINUTES • COOK TIME: 24 MINUTES • TOTAL: 34 MINUTES • SERVES: 4

Ingredients
8 tablespoons (1 stick) unsalted butter, melted
½ cup hot sauce
2 tablespoons white vinegar
2 teaspoons Worcestershire sauce

1 teaspoon garlic powder
½ cup all-purpose flour
16 frozen chicken wings

Instructions:
1 Preheat the Innsky air fryer to 370°F.
   In a small saucepan over low heat, combine the butter, hot sauce, vinegar, Worcestershire sauce, and garlic. Mix well and bring to a simmer.
   Pour the flour into a medium mixing bowl. Dredge the chicken wings in the flour.
   Place the flour-coated wings into the air fryer basket.
2   Set the timer and fry for 12 minutes. Using tongs, flip the wings.
   Reset the timer and fry for 12 minutes more.
   Release the Innsky air fryer basket from the drawer. Turn out the chicken wings into a large mixing bowl, then pour the sauce over them.
   Serve and enjoy.

Per Serving: Calories: 705; Fat: 55g; Saturated fat: 23g; Carbohydrate: 14g; Fiber: 1g; Sugar: 1g; Protein: 38g; Iron: 3mg; Sodium: 1096mg

Almond Flour Coco-Milk Battered Chicken

PREP: 5 MINUTES • COOK TIME: 30 MINUTES • TOTAL: 35 MINUTES • SERVES: 4

Ingredients

¼ cup coconut milk
½ cup almond flour
1 ½ tablespoons old bay Cajun seasoning

1 egg, beaten
4 small chicken thighs
Salt and pepper to taste

Instructions:

1   Preheat the Innsky air fryer for 5 minutes. Mix the egg and coconut milk in a bowl.
    Soak the chicken thighs in the beaten egg mixture. In a mixing bowl, combine the almond flour, Cajun seasoning, salt and pepper. Dredge the chicken thighs in the almond flour mixture. Place in the air fryer basket.

2   Cook for 30 minutes at 350°F.

Harissa-Rubbed Cornish Game Hens

PREP: 10 MINUTES PLUS 30 MINUTES TO MARINATE • COOK TIME: 20 MINUTES • TOTAL: 60 MINUTES • SERVES: 4

Ingredients

For the Harissa
½ cup olive oil
6 cloves garlic, minced
2 tablespoons smoked paprika
1 tablespoon ground coriander
1 tablespoon ground cumin
1 teaspoon ground caraway

1 teaspoon kosher salt
½ to 1 teaspoon cayenne pepper
For the Hens
½ cup yogurt
Cornish game hens, any giblets removed, split in half lengthwise

Instructions:

1   For the harissa: In a medium microwave-safe bowl, combine the oil, garlic, paprika, coriander, cumin, caraway, salt, and cayenne. Microwave on high for 1 minute, stirring halfway through the cooking time. (You can also heat this on the stovetop until the oil is hot and bubbling. Or, if you must use your air fryer for everything, cook it in the air fryer at 350°F for 5 to 6 minutes, or until the paste is heated through.)
    For the hens: In a small bowl, combine 1 to 2 tablespoons harissa and the yogurt. Whisk until well combined. Place the hen halves in a resealable plastic bag and pour the marinade over. Seal the bag and massage until all of the pieces are thoroughly coated. Marinate at room temperature for 30 minutes or in the refrigerator for up to 24 hours.

2   Arrange the hen halves in a single layer in the air fryer basket. (If you have a smaller air fryer, you may have to cook this in two batches.) Set the Innsky air fryer to 400°F for 20 minutes. Use a meat thermometer to ensure the game hens have reached an internal temperature.

Per Serving: Calories: 590; Fat: 38g; Protein:32.5g; Carbs:3.2g

Parmesan Chicken Tenders
PREP: 5 MINUTES • COOK TIME: 8 MINUTES • TOTAL: 13 MINUTES • SERVES: 4

Ingredients
1 pound chicken tenderloins

3 large egg whites

½ cup Italian-style bread crumbs

¼ cup grated Parmesan cheese

Instructions:

1. Spray the Innsky air fryer basket with olive oil. Trim off any white fat from the chicken tenders. In a small bowl, beat the egg whites until frothy. In a separate small mixing bowl, combine the bread crumbs and Parmesan cheese. Mix well. Dip the chicken tenders into the egg mixture, then into the Parmesan and bread crumbs. Shake off any excess breading. Place the chicken tenders in the greased Innsky air fryer basket in a single layer. Generously spray the chicken with olive oil to avoid powdery, uncooked breading.

2. Set the temperature of your Innsky AF to 370°F. Set the timer and bake for 4 minutes. Using tongs, flip the chicken tenders and bake for 4 minutes more. Check that the chicken has reached an internal temperature of 165°F. Add cooking time if needed. Once the chicken is fully cooked, plate, serve, and enjoy.

Per Serving: Calories: 210; Fat: 4g; Saturated fat: 1g; Carbohydrate: 10g; Fiber: 1g; Sugar: 1g; Protein: 33g; Iron: 1mg; Sodium: 390mg

Air Fryer Grilled Chicken Fajitas
PREP: 10 MINUTES • COOK TIME: 14 MINUTES • TOTAL: 24 MINUTES • SERVES: 4

Ingredients
1 pound chicken tenders

1 onion, sliced

1 yellow bell pepper, diced

1 red bell pepper, diced

1 orange bell pepper, diced

2 tablespoons olive oil

1 tablespoon fajita seasoning mix

Instructions:

1   Slice the chicken into thin strips. In a large mixing bowl, combine the chicken, onion, and peppers. Add the olive oil and fajita seasoning and mix well, so that the chicken and vegetables are thoroughly covered with oil. Place the chicken and vegetable mixture into the Innsky air fryer basket in a single layer.

2   Set the temperature of your Innsky AF to 350°F. Set the timer and grill for 7 minutes.
Shake the basket and use tongs to flip the chicken. Reset the timer and grill for 7 minutes more, or until the chicken is cooked through and the juices run clear. Once the chicken is fully cooked, transfer it to a platter and serve.

One-Dish Chicken & Rice
PREP: 10 MINUTES • COOK TIME: 40 MINUTES • TOTAL: 50 MINUTES • SERVES: 4

Ingredients
1 cup long-grain white rice

1 cup cut frozen green beans

1 tablespoon minced fresh ginger

3 cloves garlic, minced

1 tablespoon toasted sesame oil

1 teaspoon kosher salt

1 teaspoon black pepper

1pound chicken wings

Instructions:

1   In a 6 × 3-inch round heatproof pan, combine the rice, green beans, ginger, garlic, sesame oil, salt, and pepper. Stir to combine. Place the chicken wings on top of the rice mixture. Cover the pan with foil. Make a long slash in the foil to allow the pan to vent steam.

2   Place the pan in the air fryer basket. Set the Innsky air fryer to 375°F for 30 minutes. Remove the foil. Set the Innsky air fryer to 400°F for 10 minutes, or until the wings have browned and rendered fat into the rice and vegetables, turning the wings halfway through the cooking time.

Lebanese Turkey Burgers with Feta & Tzatziki
PREP: 25 MINUTES • COOK TIME: 12 MINUTES • TOTAL: 37 MINUTES • SERVES: 4

Ingredients

For the Tzatziki

1 large cucumber, peeled and grated
2 to 3 cloves garlic, minced
1 cup plain Greek yogurt
1 tablespoon tahini (sesame paste)
1 tablespoon fresh lemon juice
½ teaspoon kosher salt

For the Burgers

1 pound ground turkey, chicken, or lamb
1 small yellow onion, finely diced
1 clove garlic, minced

2 tablespoons chopped fresh parsley
2 teaspoons Lebanese Seven-Spice Mix
½ teaspoon kosher salt
Vegetable oil spray

For Serving

4 lettuce leaves or 2 whole-wheat pita breads, halved
8 slices ripe tomato
1 cup baby spinach
⅓ cup crumbled feta cheese

Instructions:

1 *For the tzatziki*: In a medium bowl, stir together all the ingredients until well combined. Cover and chill until ready to serve.

   For the burgers: In a large bowl, combine the ground turkey, onion, garlic, parsley, spice mix, and salt. Mix gently until well combined. Divide the turkey into four portions and form into round patties.

2 Spray the Innsky air fryer basket with vegetable oil spray. Place the patties in a single layer in the air fryer basket. Set the Innsky air fryer to 400°F for 12 minutes. Place one burger in each lettuce leaf or pita half. Tuck in 2 tomato slices, spinach, cheese, and some tzatziki.

Sweet And Sour Chicken
PREP: 5 MINUTES • COOK TIME: 20 MINUTES • TOTAL: 25 MINUTES • SERVES: 6

Ingredients

3 Chicken Breasts, cubed
1/2 Cup Flour
1/2 Cup Cornstarch
2 Red Peppers, sliced
1Onion,chopped
2 Carrots, julienned

3/4 Cup Sugar
2 Tbsps Cornstarch
1/3 Cup Vinegar
2/3 Cup Water
1/4 cup Soy sauce
1 Tbsp Ketchup

Instructions:

1. Preheat the Innsky air fryer to 375 degrees.
   Combine the flour, cornstarch and chicken in an air tight container and shake to combine. Remove chicken from the container and shake off any excess flour.

2. Add chicken to the Air Fryer tray and cook for 20 minutes.
   In a saucepan, whisk together sugar, water, vinegar, soy sauce and ketchup. Bring to a boil over medium heat, reduce the heat then simmer for 2 minutes. After cooking the chicken for 20 minutes, add the vegetables and sauce mixture to the Air fryer and cook for another 5 minutes. Serve over hot rice

Easy Lemon Chicken Thighs
PREP: 5 MINUTES • COOK TIME: 10 MINUTES • TOTAL: 15 MINUTES • SERVES: 4

Ingredients

1 teaspoon salt

1 teaspoon freshly ground black pepper

2 tablespoons olive oil

2 tablespoons Italian seasoning

2 tablespoons freshly squeezed lemon juice

1 lemon, sliced

Instructions:

1. Place the chicken thighs in a medium mixing bowl and season them with the salt and pepper. Add the olive oil, Italian seasoning, and lemon juice and toss until the chicken thighs are thoroughly coated with oil. Add the sliced lemons. Place the chicken thighs into the Innsky air fryer basket in a single layer.

2. Set the temperature of your Innsky AF to 350°F. Set the timer and cook for 10 minutes. Using tongs, flip the chicken. Reset the timer and cook for 10 minutes more. Check that the chicken has reached an internal temperature of 165°F. Add cooking time if needed. Once the chicken is fully cooked, plate, serve, and enjoy.

Per Serving: Calories: 325; Fat: 26g; Saturated fat: 6g; Carbohydrate: 1g; Fiber: 0g; Sugar: 1g; Protein: 20g; Iron: 1mg; Sodium: 670mg

Air Fryer Southern Fried Chicken
PREP:15 MINUTES PLUS 1HOUR TO MARINATE•COOK TIME: 26 MINUTES •TOTAL:1 HOUR 36 MINUTES•SERVES: 4

Ingredients

½ cup buttermilk

2 teaspoons salt, plus 1 tablespoon

1 teaspoon freshly ground black pepper

1 pound chicken thighs and drumsticks

1 cup all-purpose flour

2 teaspoons onion powder

2 teaspoons garlic powder

½ teaspoon sweet paprika

Instructions:

1. In a large mixing bowl, whisk together the buttermilk, 2 teaspoons of salt, and pepper. Add the chicken pieces to the bowl, and let the chicken marinate for at least an hour, covered, in the refrigerator. About 5 minutes before the chicken is done marinating, prepare the dredging mixture. In a large mixing bowl, combine the flour, 1 tablespoon of salt, onion powder, garlic powder, and paprika. Spray the Innsky air fryer basket with olive oil. Remove the chicken from the buttermilk mixture and dredge it in the flour mixture. Shake off any excess flour. Place the chicken pieces into the greased Innsky air fryer basket in a single layer, leaving space between each piece. Spray the chicken generously with olive oil.

2. Set the temperature of your Innsky AF to 390°F. Set the timer and cook for 13 minutes. Using tongs, flip the chicken. Spray generously with olive oil. Reset the timer and fry for 13 minutes more. Check that the chicken has reached an internal temperature of 165°F. Add cooking time if needed. Once the chicken is fully cooked, plate, serve, and enjoy!

Per Serving: Calories: 377; Fat: 18g; Saturated fat: 5g; Carbohydrate: 28g; Fiber: 1g; Sugar: 2g; Protein: 25g; Iron: 3mg; Sodium: 1182mg

Buffalo Chicken Wings

PREP: 5 MINUTES • COOK TIME: 30 MINUTES • TOTAL: 35 MINUTES • SERVES: 8

Ingredients

1 tsp. salt
1-2 tbsp. brown sugar
1 tbsp. Worcestershire sauce

½ C. vegan butter
½ C. cayenne pepper sauce
4 pounds chicken wings

Instructions:

1   Whisk salt, brown sugar, Worcestershire sauce, butter, and hot sauce together and set to the side. Dry wings and add to air fryer basket.

2   Set temperature to 380°F, and set time to 25 minutes. Cook tossing halfway through. When timer sounds, shake wings and bump up the temperature to 400 degrees and cook another 5 minutes.
Take out wings and place into a big bowl. Add sauce and toss well.
Serve alongside celery sticks.

Perfect Chicken Parmesan

PREP: 5 MINUTES • COOK TIME: 25 MINUTES • TOTAL: 30 MINUTES • SERVES: 2

Ingredients

2 large white meat chicken breasts
1 cup of breadcrumbs
2 medium-sized eggs
Pinch of salt and pepper

1 tablespoon of dried oregano
1 cup of marinara sauce
2 slices of provolone cheese
1 tablespoon of parmesan cheese

Instructions:

1.   Cover the basket of the Air fryer with a lining of tin foil, leaving the edges uncovered to allow air to circulate through the basket.
Preheat the Innsky air fryer to 350 degrees. In a mixing bowl, beat the eggs until fluffy and until the yolks and whites are fully combined, and set aside.
In a separate mixing bowl, combine the breadcrumbs, oregano, salt and pepper, and set aside. One by one, dip the raw chicken breasts into the bowl with dry ingredients, coating both sides; then submerge into the bowl with wet ingredients, then dip again into the dry ingredients. This double coating will ensure an extra crisp-and-delicious air-fry.
Lay the coated chicken breasts on the foil covering the Air fryer basket, in a single flat layer.

2.   Set the Innsky air fryer timer for 10 minutes.
After 10 minutes, the air fryer will turn off and the chicken should be mid-way cooked and the breaded coating starting to brown.
Using tongs, turn each piece of chicken over to ensure a full all-over fry.
ReSet the Innsky air fryer to 320 degrees for another 10 minutes.
While the chicken is cooking, pour half the marinara sauce into a 7-inch heat-safe pan.
After 15 minutes, when the air fryer shuts off, remove the fried chicken breasts using tongs and set in the marinara-covered pan. Drizzle the rest of the marinara sauce over the fried chicken, then place the slices of provolone cheese atop both of them and sprinkle the parmesan cheese over the entire pan.
ReSet the Innsky air fryer to 350 degrees for 5 minutes.
After 5 minutes, when the air fryer shuts off, remove the dish from the air fryer using tongs or oven mitts. The chicken will be perfectly crisped and the cheese melted and lightly toasted. Serve while hot!

Air Fryer Grilled Chicken Breasts
PREP: 5 MINUTES • COOK TIME: 14 MINUTES • TOTAL: 19 MINUTES • SERVES: 4

Ingredients
½ teaspoon garlic powder

1 teaspoon salt

½ teaspoon freshly ground black pepper

1 teaspoon dried parsley

2 tablespoons olive oil, divided

3 boneless, skinless chicken breasts

Instructions:
1   In a small mixing bowl, mix together the garlic powder, salt, pepper, and parsley. Using 1 tablespoon of olive oil and half of the seasoning mix, rub each chicken breast with oil and seasonings. Place the chicken breast in the air fryer basket.

2   Set the temperature of your Innsky AF to 370°F. Set the timer and grill for 7 minutes.
Using tongs, flip the chicken and brush the remaining olive oil and spices onto the chicken. Reset the timer and grill for 7 minutes more. Check that the chicken has reached an internal temperature of 165°F. Add cooking time if needed.
Once the chicken is fully cooked, transfer it to a platter and serve.

Per Serving: Calories: 182; Fat: 9g; Saturated fat: 1g; Carbohydrate: 0g; Fiber: 0g; Sugar: 0g; Protein: 26g; Iron: 1mg; Sodium: 657mg

Zingy & Nutty Chicken Wings
PREP: 5 MINUTES • COOK TIME: 18 MINUTES • TOTAL: 23 MINUTES • SERVES: 4

Ingredients
1 tablespoon fish sauce

1 tablespoon fresh lemon juice

1 teaspoon sugar

12 chicken middle wings, cut into half

2 fresh lemongrass stalks, chopped finely

¼ cup unsalted cashews, crushed

Instructions:
1.   In a bowl, mix together fish sauce, lime juice and sugar.
Add wings ad coat with mixture generously. Refrigerate to marinate for about 1-2 hours.
Preheat the Innsky air fryer to 355 degrees F.

2.   In the Air fryer pan, place lemongrass stalks. Cook for about 2-3 minutes. Remove the cashew mixture from Air fryer and transfer into a bowl. Now, Set the Innsky air fryer to 390 degrees F.
Place the chicken wings in Air fryer pan. Cook for about 13-15 minutes further.
Transfer the wings into serving plates. Sprinkle with cashew mixture and serve.

Pesto-Cream Chicken with Cherry Tomatoes
PREP: 10 MINUTES • COOK TIME: 15 MINUTES • TOTAL: 25 MINUTES • SERVES: 4

Ingredients
Vegetable oil spray

½ cup prepared pesto

¼ cup half-and-half

¼ grated Parmesan cheese

½ to 1 teaspoon red pepper flakes

1 pound boneless, skinless chicken thighs, halved crosswise

1 small onion, sliced

½cup sliced red and/or green bell peppers

½ cup halved cherry tomatoes

Instructions:
1   Spray a 6 × 3-inch round heatproof pan with vegetable oil spray; set aside. In a large bowl, combine the pesto, half-and-half, cheese, and red pepper flakes. Whisk until well combined. Add the chicken and turn to coat. Transfer the sauce and chicken to the prepared pan. Scatter the onion, bell pepper, and tomatoes on top.

2   Place the pan in the air fryer basket. Set the Innsky air fryer to 350°F for 15 minutes. Use a meat thermometer to ensure the chicken has reached an internal temperature of 165°F.

Peanut Chicken

PREP: 15 MINUTES PLUS 30 MINUTES TO MARINATE • COOK TIME: 20 MINUTES • TOTAL: 1 HOUR 5 MINUTES • SERVES: 4

Ingredients

¼ cup creamy peanut butter
2 tablespoons sweet chili sauce
2 tablespoons fresh lime juice
1 tablespoon sriracha
1 tablespoon soy sauce
1 teaspoon minced fresh ginger
1 clove garlic, minced

½ teaspoon kosher salt
½ cup hot water
1 pound bone-in chicken thighs
2 tablespoons chopped fresh cilantro, for garnish
¼ cup chopped green onions, for garnish
2 to 3 tablespoons crushed roasted and salted peanuts, for garnish

Instructions:

1 In a small bowl, combine the peanut butter, sweet chili sauce, lime juice, sriracha, soy sauce, ginger, garlic, and salt. Add the hot water and whisk until smooth. Place the chicken in a resealable plastic bag and pour in half of the sauce. Reserve the remaining sauce for serving. Seal the bag and massage until all of the chicken is well coated. Marinate at room temperature for 30 minutes or in the refrigerator for up to 24 hours. Remove the chicken from the bag and discard the marinade. Place the chicken in the air fryer basket.

2 Set the Innsky air fryer to 350°F for 20 minutes. Use a meat thermometer to ensure the chicken has reached an internal temperature of 165°F. Transfer the chicken to a serving platter. Sprinkle with the cilantro, green onions, and peanuts. Serve with the reserved sauce for dipping.

Basil-Garlic Breaded Chicken Bake

PREP: 5 MINUTES • COOK TIME: 25 MINUTES • TOTAL: 30 MINUTES • SERVES: 2

Ingredients

2 boneless skinless chicken breast halves
1 tablespoon butter, melted
1 large tomato, seeded and chopped
2 garlic cloves, minced
1 1/2 tablespoons minced fresh basil
1/2 tablespoon olive oil

1/2 teaspoon salt
1/4 cup all-purpose flour
1/4 cup egg substitute
1/4 cup grated Parmesan cheese
1/4 cup dry bread crumbs
1/4 teaspoon pepper

Instructions:

1 In shallow bowl, whisk well egg substitute and place flour in a separate bowl. Dip chicken in flour, then egg, and then flour. In small bowl whisk well butter, bread crumbs and cheese. Sprinkle over chicken. Lightly grease baking pan of air fryer with cooking spray. Place breaded chicken on bottom of pan. Cover with foil.

2 For 20 minutes, cook on 390°F.
Meanwhile, in a bowl whisk well remaining ingredient. Remove foil from pan and then pour over chicken the remaining Ingredients. Cook for 8 minutes. Serve and enjoy.

Per Serving: Calories: 311; Fat: 11g; Protein:31g; Carbs:22g

Crispy Air Fryer Butter Chicken
PREP: 5 MINUTES • COOK TIME: 14 MINUTES • TOTAL: 19 MINUTES • SERVES: 2

Ingredients

2 (8-ounce) boneless, skinless chicken breasts
1 sleeve Ritz crackers

4 tablespoons (½ stick) cold unsalted butter, cut into 1-tablespoon slices

Instructions:

1   Spray the Innsky air fryer basket with olive oil, or spray an air fryer–size baking sheet with olive oil or cooking spray.

Dip the chicken breasts in water. Put the crackers in a resealable plastic bag. Using a mallet or your hands, crush the crackers. Place the chicken breasts inside the bag one at a time and coat them with the cracker crumbs.

Place the chicken in the greased air fryer basket, or on the greased baking sheet set into the air fryer basket. Put 1 to 2 dabs of butter onto each piece of chicken.

2   Set the temperature of your Innsky AF to 370°F. Set the timer and bake for 7 minutes.

Using tongs, flip the chicken. Spray the chicken generously with olive oil to avoid uncooked breading. Reset the timer and bake for 7 minutes more.

Check that the chicken has reached an internal temperature of 165°F. Add cooking time if needed. Using tongs, remove the chicken from the air fryer and serve.

Per Serving: Calories: 750; Fat: 40g; Saturated fat: 18g; Carbohydrate: 38g; Fiber: 2g; Sugar: 5g; Protein: 57g; Iron: 4mg; Sodium: 853mg

South Indian Pepper Chicken
PREP:20 PLUS 30 MINUTES TO MARINATE•COOK TIME:15 MINUTES•TOTAL:1 HOUR 5 MINUTES •SERVES: 4

Ingredients

For the Spice Mix
1 dried red chile, or ½ teaspoon dried red pepper flakes
1-inch piece cinnamon or cassia bark
1½ teaspoons coriander seeds
1 teaspoon fennel seeds
1 teaspoon cumin seeds
1 teaspoon black peppercorns
½ teaspoon cardamom seeds

¼ teaspoon ground turmeric
1 teaspoon kosher salt
For the Chicken
1pound boneless, skinless chicken thighs, cut crosswise into thirds
2 medium onions, cut into ½-inch-thick slices
¼ cup olive oil
Cauliflower rice, steamed rice, or naan bread, for serving

Instructions:

1   For the spice mix: Combine the dried chile, cinnamon, coriander, fennel, cumin, peppercorns, and cardamom in a clean coffee or spice grinder. Grind, shaking the grinder lightly so all the seeds and bits get into the blades, until the mixture is broken down to a fine powder. Stir in the turmeric and salt. For the chicken: Place the chicken and onions in resealable plastic bag. Add the oil and 1½ tablespoons of the spice mix. Seal the bag and massage until the chicken is well coated. Marinate at room temperature for 30 minutes or in the refrigerator for up to 24 hours.

2   Place the chicken and onions in the air fryer basket. Set the Innsky air fryer to 350°F for 10 minutes, stirring once halfway through the cooking time. Increase the temperature to 400°F for 5 minutes. Use a meat thermometer to ensure the chicken has reached an internal temperature of 165°F.Serve with steamed rice, cauliflower rice, or naan.

Honey and Wine Chicken Breasts
PREP: 5 MINUTES • COOK TIME: 15 MINUTES • TOTAL: 20 MINUTES • SERVES: 4

Ingredients
2 chicken breasts, rinsed and halved

1 tablespoon melted butter

1/2 teaspoon freshly ground pepper, or to taste

3/4 teaspoon sea salt, or to taste

1 teaspoon paprika

1 teaspoon dried rosemary

2 tablespoons dry white wine

1 tablespoon honey

Instructions:
1   Firstly, pat the chicken breasts dry. Lightly coat them with the melted butter. Then, add the remaining ingredients.

2   Transfer them to the air fryer basket; bake about 15 minutes at 330 degrees F. Serve warm and enjoy

Light And Airy Breaded Chicken Breasts
PREP: 5 MINUTES • COOK TIME: 14 MINUTES • TOTAL: 19 MINUTES • SERVES: 2

Ingredients
2 large eggs

1cup bread crumbs or panko bread crumbs

1 teaspoon Italian seasoning

4 to 5 tablespoons vegetable oil

2 boneless, skinless, chicken breasts

Instructions:
1   Preheat the Innsky air fryer to 370°F. Spray the Innsky air fryer basket with olive oil or cooking spray. In a small mixing bowl, beat the eggs until frothy. In a separate small mixing bowl, mix together the bread crumbs, Italian seasoning, and oil. Dip the chicken in the egg mixture, then in the bread crumb mixture. Place the chicken directly into the greased air fryer basket, or on the greased baking sheet set into the basket.

2   Spray the chicken generously and thoroughly with olive oil to avoi powdery, uncooked breading. Set the timer and fry for 7 minutes. Using tongs, flip the chicken and generously spray it with olive oil. Reset the timer and fry for 7 minutes more. Check that the chicken has reached an internal temperature of 165°F. Add cooking time if needed. Once the chicken is fully cooked, use tongs to remove it from the air fryer and serve.

Per Serving: Calories: 833; Fat: 46g; Saturated fat: 5g; Carbohydrate: 40g; Fiber: 2g; Sugar: 4g; Protein: 65g; Iron: 5mg; Sodium: 609mg

Air Fryer Chicken Parmesan
PREP: 5 MINUTES • COOK TIME: 14 MINUTES • TOTAL: 19 MINUTES • SERVES: 4

Ingredients
2 boneless, skinless chicken breasts

2 large eggs

1 cup Italian-style bread crumbs

¼ cup shredded Parmesan cheese

½ cup marinara sauce

½ cup shredded mozzarella cheese

Instructions:
1   Preheat the Innsky air fryer to 360°F. Spray an air fryer–size baking sheet with olive oil or cooking spray. Using a mallet or rolling pin, flatten the chicken breasts to about ¼ inch thick. In a small mixing bowl, beat the eggs until frothy. In another small mixing bowl, mix together the bread crumbs and Parmesan cheese. Dip the chicken in the egg, then in the bread crumb mixture. Place the chicken on the greased baking sheet. Set the baking sheet into the air fryer basket. Spray the chicken generously with olive oil to avoid powdery, uncooked breading.

2   Set the timer and bake for 7 minutes, or until cooked through and the juices run clear. Flip the chicken and pour the marinara sauce over the chicken. Sprinkle with the mozzarella cheese. Reset the timer and bake for another 7 minutes.

Spicy Roast Chicken

PREP: 10 MINUTES PLUS 30 MINUTES TO MARINATE • COOK TIME: 10 MINUTES • TOTAL: 50 MINUTES • SERVES: 4

Ingredients

1 teaspoon ground turmeric
½ teaspoon cayenne pepper
½ teaspoon ground cinnamon
¼ teaspoon ground cloves
¼ teaspoon kosher salt
1 tablespoon cider vinegar

2 tablespoons olive oil
1 pound boneless, skinless chicken thighs, cut crosswise into thirds
Zoodles, steamed rice, naan bread, or a mixed salad, for serving

Instructions:

1. In a small bowl, combine the turmeric, cayenne, cinnamon, cloves, salt, vinegar, and oil. Stir to form a thick paste. Place the chicken in a resealable plastic bag and add the marinade. Seal the bag and massage until the chicken is well coated. Marinate at room temperature for 30 minutes or in the refrigerator for up to 24 hours.
2. Place the chicken in the air fryer basket. Set the Innsky air fryer to 350°F for 10 minutes, turning the chicken halfway through the cooking time. Use a meat thermometer to ensure that the chicken has reached an internal temperature of 165°F.
   Serve with steamed rice or naan, over zoodles, or with a mixed salad.

Air Fryer Chicken Drumsticks With A Sweet Rub
PREP: 5 MINUTES • COOK TIME: 20 MINUTES • TOTAL: 25 MINUTES • SERVES: 4

Ingredients

¼ cup brown sugar
1 tablespoon salt
½ teaspoon freshly ground black pepper
1 teaspoon chili powder
1 teaspoon smoked paprika

1 teaspoon dry mustard
1 teaspoon garlic powder
1 teaspoon onion powder
4 to 6 chicken drumsticks
2 tablespoons olive oil

Instructions:

1. In a small mixing bowl, combine the brown sugar, salt, pepper, chili powder, paprika, mustard, garlic powder, and onion powder.
   Using a paper towel, wipe any moisture off the chicken. Put the chicken drumsticks into a large resealable plastic bag, then pour in the dry rub. Seal the bag. Shake the bag to coat the chicken. Place the drumsticks in the air fryer basket. Brush the drumsticks with olive oil.
2. Set the temperature of your Innsky AF to 390°F. Set the timer and bake for 10 minutes.
   Using tongs, flip the drumsticks, and brush them with olive oil. Reset the timer and bake for 10 minutes more. Check that the chicken has reached an internal temperature of 165°F. Add cooking time if needed. Once the chicken is fully cooked, transfer it to a platter and serve.

Chicken Fillets, Brie & Ham
PREP: 5 MINUTES • COOK TIME: 15 MINUTES • TOTAL: 20 MINUTES • SERVES: 4

Ingredients

2 Large Chicken Fillets
Freshly Ground Black Pepper
4 Small Slices of Brie (Or your cheese of choice)

1 Tbsp Freshly Chopped Chives
4 Slices Cured Ha

Instructions:

1. Slice the fillets into four and make incisions as you would for a hamburger bun. Leave a little "hinge" uncut at the back. Season the inside and pop some brie and chives in there. Close them, and wrap them each in a slice of ham. Brush with oil and pop them into the basket.
2. Heat your fryer to 350° F. Roast the little parcels until they look tasty (15 min)

Air Fried Turkey Breast

PREP: 5 MINUTES • COOK TIME: 45 MINUTES • TOTAL: 50 MINUTES • SERVES: 4

Ingredients

2 tablespoons unsalted butter

1 teaspoon salt

½ teaspoon freshly ground black pepper

1 teaspoon dried thyme

1 teaspoon dried oregano

1 (3½-pound) boneless turkey breast

1 tablespoon olive oil

Instructions:

1   Melt the butter in a small microwave-safe bowl on low for about 45 seconds. Add the salt, pepper, thyme, and oregano to the melted butter. Let the butter cool until you can handle it without burning yourself. Rub the butter mixture all over the turkey breast, then rub on the olive oil, over the butter. Place the turkey breast in the air fryer basket, skin-side down.

2   Set the temperature of your Innsky AF to 350°F. Set the timer and roast for 20 minutes.
Using tongs, flip the turkey. Reset the timer and roast the turkey breast for another 30 minutes. Check that it has reached an internal temperature of 165°F. Add cooking time if needed.
Using tongs, remove the turkey from the air fryer and let rest for about 10 minutes before carving.

Per Serving: Calories: 484; Fat: 11g; Saturated fat: 4g; Carbohydrate: 1g; Fiber: 0g; Sugar: 0g; Protein: 95g; Iron: 1mg; Sodium: 623mg

Air Fryer Cornish Hen

PREP: 5 MINUTES • COOK TIME: 30 MINUTES • TOTAL: 35 MINUTES • SERVES: 2

Ingredients

2 tablespoons Montreal chicken seasoning

1 (1½- to 2-pound) Cornish hen

Instructions:

1   Preheat the Innsky air fryer to 390°F. Rub the seasoning over the chicken, coating it thoroughly.

2   Place the chicken in the air fryer basket. Set the timer and roast for 15 minutes.
Flip the chicken and cook for another 15 minutes. Check that the chicken has reached an internal temperature of 165°F. Add cooking time if needed.

Per Serving: Calories: 520; Fat: 36g; Saturated fat: 10g; Carbohydrate: 0g; Fiber: 0g; Sugar: 0g; Protein: 45g; Iron: 2mg; Sodium: 758mg

Chicken Fajitas

PREP: 10 MINUTES • COOK TIME: 10 MINUTES • TOTAL: 20 MINUTES • SERVES: 4

Ingredients

4 boneless, skinless chicken breasts, sliced

1 small red onion, sliced

2 red bell peppers, sliced

½ cup spicy ranch salad dressing, divided

½ teaspoon dried oregano

8 corn tortillas

2 cups torn butter lettuce

2 avocados, peeled and chopped

Instructions:

1.   Place the chicken, onion, and pepper in the Air fryer basket. Drizzle with 1 tablespoon of the salad dressing and add the oregano. Toss to combine.

2.   Grill for 10 to 14 minutes or until the chicken is 165°F on a food thermometer. Transfer the chicken and vegetables to a bowl and toss with the remaining salad dressing. Serve the chicken mixture with the tortillas, lettuce, and avocados and let everyone make their own creations.

Per Serving: Calories: 783; Fat: 38g; Protein:72; Fiber:12g

Air Fried Turkey Wings

PREP: 5 MINUTES • COOK TIME: 26 MINUTES • TOTAL: 15 MINUTES • SERVES: 4

Ingredients

2 pounds turkey wings
3 tablespoons olive oil or sesame oil

3 to 4 tablespoons chicken rub

Instructions:

1   Put the turkey wings in a large mixing bowl. Pour the olive oil into the bowl and add the rub. Using your hands, rub the oil mixture over the turkey wings. Place the turkey wings in the air fryer basket.

2   Set the temperature of your Innsky AF to 380°F. Set the timer and roast for 13 minutes. Using tongs, flip the wings. Reset the timer and roast for 13 minutes more. Remove the turkey wings from the air fryer, plate, and serve.

Per Serving: Calories: 521; Fat: 34g; Saturated fat: 2g; Carbohydrate: 4g; Fiber: 0g; Sugar: 0g; Protein: 52g; Iron: 0mg; Sodium: 600mg

Crispy Honey Garlic Chicken Wings

PREP: 10 MINUTES • COOK TIME: 25 MINUTES • TOTAL: 35 MINUTES • SERVES: 8

Ingredients

1/8 C. water
½ tsp. salt
4 tbsp. minced garlic
¼ C. vegan butter

¼ C. raw honey
¾ C. almond flour
16 chicken wings

Instructions:

1   Rinse off and dry chicken wings well. Spray Innsky air fryer basket with olive oil. Coat chicken wings with almond flour and add coated wings to air fryer.

2   Set temperature to 380°F, and set time to 25 minutes. Cook shaking every 5 minutes. When the timer goes off, cook 5-10 minutes at 400 degrees till skin becomes crispy and dry. As chicken cooks, melt butter in a saucepan and add garlic. Sauté garlic 5 minutes. Add salt and honey, simmering 20 minutes. Make sure to stir every so often, so the sauce does not burn. Add a bit of water after 15 minutes to ensure sauce does not harden. Take out chicken wings from air fryer and coat in sauce. Enjoy.

BBQ Chicken Recipe from Greece

PREP: 10 MINUTES • COOK TIME: 25 MINUTES • TOTAL: 35 MINUTES • SERVES: 4

Ingredients

1 (8 ounce) container fat-free plain yogurt
2 tablespoons fresh lemon juice
2 teaspoons dried oregano
1-pound skinless, boneless chicken breast halves - cut into 1-inch pieces
1 large red onion, cut into wedges
1/2 teaspoon lemon zest

1/2 teaspoon salt
1 large green bell pepper, cut into 1 1/2-inch pieces
1/3 cup crumbled feta cheese with basil and sun-dried tomatoes
1/4 teaspoon ground black pepper
1/4 teaspoon crushed dried rosemary

Instructions:

1   In a shallow dish, mix well rosemary, pepper, salt, oregano, lemon juice, lemon zest, feta cheese, and yogurt. Add chicken and toss well to coat. Marinate in the ref for 3 hours. Thread bell pepper, onion, and chicken pieces in skewers. Place on skewer rack.

2   For 12 minutes, cook on 360°F. Halfway through cooking time, turnover skewers. If needed, cook in batches. Serve and enjoy.

Per Serving: Calories: 242; Fat: 7.5g; Protein:31g; Sugar:6g

Cheesy Chicken in Leek-Tomato Sauce
PREP: 10 MINUTES • COOK TIME: 20 MINUTES • TOTAL: 30 MINUTES • SERVES: 4

Ingredients
2 large-sized chicken breasts, cut in half lengthwise
Salt and ground black pepper, to taste
4 ounces Cheddar cheese, cut into sticks
1 tablespoon sesame oil
1 cup leeks, chopped

2 cloves garlic, minced
2/3 cup roasted vegetable stock
2/3 cup tomato puree
1 teaspoon dried rosemary
1 teaspoon dried thyme

Instructions:
Firstly, season chicken breasts with the salt and black pepper; place a piece of Cheddar cheese in the middle. Then, tie it using a kitchen string; drizzle with sesame oil and reserve. Add the leeks and garlic to the oven safe bowl. Cook in the Air fryer at 390 degrees F for 5 minutes or until tender.

Add the reserved chicken. Throw in the other ingredients and cook for 12 to 13 minutes more or until the chicken is done. Enjoy.

Chicken Kiev
PREP: 15 MINUTES • COOK TIME: 25 MINUTES • TOTAL: 40 MINUTES • SERVES: 4

Ingredients
1 cup (2 sticks) unsalted butter, softened
2 tablespoons lemon juice
2 tablespoons plus 1 teaspoon chopped fresh parsley leaves, divided, plus more for garnish
2 tablespoons chopped fresh tarragon leaves
3 cloves garlic, minced
1 teaspoon fine sea salt, divided

4 boneless, skinless chicken breasts
2 large eggs
2 cups pork dust
1 teaspoon ground black pepper
Sprig of fresh parsley, for garnish
Lemon slices, for serving

Instructions:
1   Spray the Innsky air fryer basket with avocado oil. Preheat the Innsky air fryer to 350°F. In a medium-sized bowl, combine the butter, lemon juice, 2 tablespoons of the parsley, the tarragon, garlic, and ¼ teaspoon of the salt. Cover and place in the fridge to harden for 7 minutes. While the butter mixture chills, place one of the chicken breasts on a cutting board. With a sharp knife held parallel to the cutting board, make a 1-inch-wide incision at the top of the breast. Carefully cut into the breast to form a large pocket, leaving a ½-inch border along the sides and bottom. Repeat with the other 3 breasts. Stuff one-quarter of the butter mixture into each chicken breast and secure the openings with toothpicks. Beat the eggs in a small shallow dish. In another shallow dish, combine the pork dust, the remaining 1 teaspoon of parsley, the remaining ¾ teaspoon of salt, and the pepper. One at a time, dip the chicken breasts in the egg, shake off the excess egg, and dredge the breasts in the pork dust mixture. Use your hands to press the pork dust onto each breast to form a nice crust. If you desire a thicker coating, dip it again in the egg and pork dust. As you finish, spray each coated chicken breast with avocado oil and place it in the air fryer basket.

2   Cook the chicken in the air fryer for 15 minutes, flip the breasts, and cook for another 10 minutes, or until the internal temperature of the chicken is 165°F and the crust is golden brown.
Serve garnished with chopped fresh parsley and a parsley sprig, with lemon slices on the side.
Store leftovers in an airtight container in the refrigerator for up to 4 days or in the freezer for up to a month. Reheat in a preheated 350°F air fryer for 5 minutes, or until heated through.

Per serving: Calories 801; Fat 64g; Protein 51g; Total carbs 3g; Fiber 1g

Chicken-Fried Steak Supreme
PREP: 10 MINUTES • COOK TIME: 30 MINUTES • TOTAL: 40 MINUTES • SERVES: 8

Ingredients

½ pound beef-bottom round, sliced into strips
1 cup of breadcrumbs
2 medium-sized eggs

Pinch of salt and pepper
½ tablespoon of ground thyme

Instructions:

1. Cover the basket of the Air fryer with a lining of tin foil, leaving the edges uncovered to allow air to circulate through the basket. Preheat the Innsky air fryer to 350 degrees. In a mixing bowl, beat the eggs until fluffy and until the yolks and whites are fully combined, and set aside. In a separate mixing bowl, combine the breadcrumbs, thyme, salt and pepper, and set aside. One by one, dip each piece of raw steak into the bowl with dry ingredients, coating all sides; then submerge into the bowl with wet ingredients, then dip again into the dry ingredients. This double coating will ensure an extra crisp air fry. Lay the coated steak pieces on the foil covering the air-fryer basket, in a single flat layer.
2. Set the Innsky air fryer timer for 15 minutes. After 15 minutes, the air fryer will turn off and the steak should be mid-way cooked and the breaded coating starting to brown. Using tongs, turn each piece of steak over to ensure a full all-over fry. ReSet the Innsky air fryer to 320 degrees for 15 minutes. After 15 minutes, when the air fryer shuts off, remove the fried steak strips using tongs and set on a serving plate. Eat as soon as cool enough to handle and enjoy.

Caesar Marinated Grilled Chicken
PREP: 10 MINUTES • COOK TIME: 24 MINUTES • TOTAL: 34 MINUTES • SERVES: 3

Ingredients

¼ cup crouton
1 teaspoon lemon zest. Form into ovals, skewer and grill.
1/2 cup Parmesan

1/4 cup breadcrumbs
1-pound ground chicken
2 tablespoons Caesar dressing and more for drizzling
2-4 romaine leaves

Instructions:

1. In a shallow dish, mix well chicken, 2 tablespoons Caesar dressing, parmesan, and breadcrumbs. Mix well with hands. Form into 1-inch oval patties. Thread chicken pieces in skewers. Place on skewer rack in air fryer.
2. For 12 minutes, cook on 360°F. Halfway through cooking time, turnover skewers. If needed, cook in batches. Serve and enjoy on a bed of lettuce and sprinkle with croutons and extra dressing.

Fried Chicken Livers
PREP: 5 MINUTES • COOK TIME: 10 MINUTES • TOTAL: 15 MINUTES • SERVES: 4

Ingredients

1 pound chicken livers
1 cup flour
1/2 cup cornmeal
2 teaspoons your favorite seasoning blend

3 eggs
2 tablespoons milk

Instructions:

1. Clean and rinse the livers, pat dry. Beat eggs in a shallow bowl and mix in milk. In another bowl combine flour, cornmeal, and seasoning, mixing until even.
   Dip the livers in the egg mix, then toss them in the flour mix.
2. Air-fry at 375 degrees for 10 minutes using your Air fryer. Toss at least once halfway through.

Easy Thanksgiving Turkey Breast

PREP: 5 MINUTES • COOK TIME: 30 MINUTES • TOTAL: 35 MINUTES • SERVES: 4

Ingredients

1½ teaspoons fine sea salt
1 teaspoon ground black pepper
1 teaspoon chopped fresh rosemary leaves
1 teaspoon chopped fresh sage
1 teaspoon chopped fresh tarragon

1 teaspoon chopped fresh thyme leaves
1 (2-pound) turkey breast
3 tablespoons ghee or unsalted butter, melted
3 tablespoons Dijon mustard

Instructions:

1   Spray the air fryer with avocado oil. Preheat the Innsky air fryer to 390°F. In a small bowl, stir together the salt, pepper, and herbs until well combined. Season the turkey breast generously on all sides with the seasoning. In another small bowl, stir together the ghee and Dijon. Brush the ghee mixture on all sides of the turkey breast.

2   Place the turkey breast in the Innsky air fryer basket and cook for 30 minutes, or until the internal temperature reaches 165°F. Transfer the breast to a cutting board and allow it to rest for 10 minutes before cutting it into ½-inch-thick slices.

Store leftovers in an airtight container in the refrigerator for up to 4 days or in the freezer for up to a month. Reheat in a preheated 350°F air fryer for 4 minutes, or until warmed through.

Per serving: Calories 388; Fat 18g; Protein 50g; Total carbs 1g; Fiber 0.3g

Chicken Cordon Bleu Meatballs

PREP: 10 MINUTES • COOK TIME: 15 MINUTES • TOTAL: 25 MINUTES • SERVES: 4

Ingredients

MEATBALLS:
½ pound ground chicken
½ pound ham, diced
½ cup finely grated Swiss cheese (about 2 ounces)
¼ cup chopped onions
3 cloves garlic, minced
1½ teaspoons fine sea salt
1 teaspoon ground black pepper, plus more for garnish if desired

1 large egg, beaten
DIJON SAUCE:
¼ cup chicken broth, hot
3 tablespoons Dijon mustard
2 tablespoons lemon juice
¾ teaspoon fine sea salt
¼ teaspoon ground black pepper
Chopped fresh thyme leaves, for garnish (optional)

Instructions:

1   **Spray the Innsky air fryer basket with avocado oil. Preheat the Innsky air fryer to 390°F.**
In a large bowl, mix all the ingredients for the meatballs with your hands until well combined. Shape the meat mixture into about twelve 1½-inch balls.

2

Place the meatballs in the air fryer basket, leaving space between them, and cook for 15 minutes, or until cooked through and the internal temperature reaches 165°F.

While the meatballs cook, make the sauce: In a small mixing bowl, stir together all the sauce ingredients until well combined. Pour the sauce into a serving dish and place the meatballs on top. Garnish with ground black pepper and fresh thyme leaves, if desired.

Store leftover meatballs in an airtight container in the refrigerator for up to 5 days or in the freezer for up to a month. Reheat in a preheated 350°F air fryer for 4 minutes, or until heated through.

Per serving: Calories 288; Fat 15g; Protein 31g; Total carbs 5g; Fiber 0.5g

Cheesy Chicken Tenders

PREP: 10 MINUTES • COOK TIME: 30 MINUTES • TOTAL: 40 MINUTES • SERVES: 4

Ingredients

1 large white meat chicken breast
1 cup of breadcrumbs
2 medium-sized eggs

Pinch of salt and pepper
1 tablespoon of grated or powdered parmesan cheese

Instructions:

1. Cover the basket of the Air fryer with a lining of tin foil, leaving the edges uncovered to allow air to circulate through the basket. Preheat the Innsky air fryer to 350 degrees. In a mixing bowl, beat the eggs until fluffy and until the yolks and whites are fully combined, and set aside. In a separate mixing bowl, combine the breadcrumbs, parmesan, salt and pepper, and set aside. One by one, dip each piece of raw chicken into the bowl with dry ingredients, coating all sides; then submerge into the bowl with wet ingredients, then dip again into the dry ingredients. Lay the coated chicken pieces on the foil covering the Air fryer basket, in a single flat layer.

2. Set the Innsky air fryer timer for 15 minutes. After 15 minutes, the air fryer will turn off and the chicken should be mid-way cooked and the breaded coating starting to brown. Using tongs turn each piece of chicken over to ensure a full all over fry. ReSet the Innsky air fryer to 320 degrees for another 15 minutes. After 15 minutes, when the air fryer shuts off, remove the fried chicken strips using tongs and set on a serving plate. Eat as soon as cool enough to handle, and enjoy.

Per Serving: Calories: 278; Fat: 15g; Protein:29g; Sugar:7g

Buffalo Chicken Drumsticks

PREP: 10 MINUTES PLUS 2 HOURS TO MARINATE • COOK TIME: 25 MINUTES • TOTAL: 2 HOURS 35 MINUTES •

SERVES: 8

Ingredients

1 cup dill pickle juice
2 pounds chicken drumsticks
2 teaspoons fine sea salt
WING SAUCE:
⅓ cup hot sauce
¼ cup unsalted butter, melted
1 tablespoon lime juice

½ teaspoon fine sea salt
⅛ teaspoon garlic powder
FOR SERVING:
½ cup Blue Cheese Dressing or Ranch Dressing
Celery sticks

Instructions:

1 Place the dill pickle juice in a large shallow dish and add the chicken. Spoon the juice over the chicken, cover, and place in the fridge to marinate for 2 hours or overnight.
Spray the Innsky air fryer basket with avocado oil. Preheat the Innsky air fryer to 400°F.

2
Pat the chicken dry and season it well with the salt. Cook in the air fryer for 20 minutes, or until the internal temperature reaches 165°F, flipping after 15 minutes.
While the chicken cooks, make the wing sauce: In a large mixing bowl, stir together all the sauce ingredients until well combined.
Remove the drumsticks from the air fryer and place them in the bowl with the sauce. Coat the drumsticks well with the sauce, then use tongs or a slotted spoon to return them to the Innsky air fryer basket and cook for 5 minutes more. Serve with any extra wing sauce, blue cheese dressing, and celery sticks. Store extra drumsticks in an airtight container in the fridge for up to 4 days or in the freezer for up to a month. Reheat in a preheated 350°F air fryer for 5 minutes, then increase the temperature to 400°F and cook for 3 to 5 minutes more, until warm and crispy.

Per serving; Calories 472; Fat 34g; Protein 38g; Total carbs 1g; Fiber 0.3g

Lemon-Pepper Chicken Wings
PREP: 10 MINUTES • COOK TIME: 20 MINUTES • TOTAL: 30 MINUTES • SERVES: 4

Ingredients

8 whole chicken wings
Juice of ½ lemon
½ teaspoon garlic powder
1 teaspoon onion powder
Salt

Pepper
¼ cup low-fat buttermilk
½ cup all-purpose flour
Cooking oil

Instructions:

1. Place the wings in a sealable plastic bag. Drizzle the wings with the lemon juice. Season the wings with the garlic powder, onion powder, and salt and pepper to taste. Seal the bag. Shake thoroughly to combine the seasonings and coat the wings. Pour the buttermilk and the flour into separate bowls large enough to dip the wings. Spray the Innsky air fryer basket with cooking oil.
   One at a time, dip the wings in the buttermilk and then the flour.

2. Place the wings in the Air fryer basket. It is okay to stack them on top of each other. Spray the wings with cooking oil, being sure to spray the bottom layer. Cook for 5 minutes. Remove the basket and shake it to ensure all of the pieces will cook fully.
   Return the basket to the Air fryer and continue to cook the chicken. Repeat shaking every 5 minutes until a total of 20 minutes has passed. Cool before serving.

Per Serving: Calories: 347; Fat: 12g; Protein:46g; Fiber:1g

Mexican Chicken Burgers
PREP: 10 MINUTES • COOK TIME: 10 MINUTES • TOTAL: 20 MINUTES • SERVES: 6

Ingredients

1 jalapeno pepper
1 tsp. cayenne pepper
1 tbsp. mustard powder
1 tbsp. oregano
1 tbsp. thyme

3 tbsp. smoked paprika
1 beaten egg
1 small head of cauliflower
4 chicken breasts

Instructions:

1. Ensure your Air fryer is preheated to 350 degrees.
   Add seasonings to a blender. Slice cauliflower into florets and add to blender. Pulse till mixture resembles that of breadcrumbs. Take out ¾ of cauliflower mixture and add to a bowl. Set to the side. In another bowl, beat your egg and set to the side. Remove skin and bones from chicken breasts and add to blender with remaining cauliflower mixture. Season with pepper and salt. Take out mixture and form into burger shapes. Roll each patty in cauliflower crumbs, then the egg, and back into crumbs again.

2. Place coated patties into the air fryer. Set temperature to 350°F, and set time to 10 minutes. Flip over at 10-minute mark. They are done when crispy!

Per Serving: Calories: 234; Fat: 18g; Protein:24g; Sugar:1g

Sesame Turkey Balls in Lettuce Cups
PREP: 10 MINUTES • COOK TIME: 15 MINUTES • TOTAL: 25 MINUTES • SERVES: 6

Ingredients

MEATBALLS:

2 pounds ground turkey

2 large eggs, beaten

¾ cup finely chopped button mushrooms

¼ cup finely chopped green onions, plus more for garnish if desired

2 tablespoons Swerve confectioners'-style sweetener

2 teaspoons peeled and grated fresh ginger

2 teaspoons toasted sesame oil

1½ teaspoons wheat-free tamari, or 2 tablespoons coconut aminos

1 clove garlic, smashed to a paste

SAUCE:

½ cup chicken broth

⅓ cup Swerve confectioners'-style sweetener

2 tablespoons toasted sesame oil

2 tablespoons tomato sauce

2 tablespoons wheat-free tamari, or ½ cup coconut aminos

1 tablespoon lime juice

¼ teaspoon peeled and grated fresh ginger

1 clove garlic, smashed to a paste

Boston lettuce leaves, for serving

Sliced red chiles, for garnish (optional)

Toasted sesame seeds, for garnish (optional)

Instructions:

1 Preheat the Innsky air fryer to 350°F. Place all the ingredients for the meatballs in a large bowl and, using your hands, mix them together until well combined. Shape the mixture into about twelve 1½-inch meatballs and place them in a pie pan that will fit in the air fryer, leaving space between them. Make the sauce: In a medium-sized bowl, stir together all the sauce ingredients until well combined. Pour the sauce over the meatballs.

2 Place the pan in the air fryer and cook for 15 minutes, or until the internal temperature of the meatballs reaches 165°F, flipping after 6 minutes. To serve, lay several lettuce leaves on a serving plate and place several meatballs on top. Garnish with sliced red chiles, green onions, and/or sesame seeds, if desired.

Per serving: Calories 322; Fat 19g; Protein 32g; Total carbs 2g; Fiber 0.3g

Porchetta-Style Chicken Breasts
PREP: 10 MINUTES • COOK TIME: 15 MINUTES • TOTAL: 25 MINUTES • SERVES: 4

Ingredients

½ cup fresh parsley leaves

¼ cup roughly chopped fresh chives

4 cloves garlic, peeled

2 tablespoons lemon juice

3 teaspoons fine sea salt

1 teaspoon dried rubbed sage

1 teaspoon fresh rosemary leaves

1 teaspoon ground fennel

½ teaspoon red pepper flakes

4 (4-ounce) boneless, skinless chicken breasts, pounded to ¼ inch thick

8 slices bacon

Sprigs of fresh rosemary, for garnish (optional)

Instructions:

1 Spray the Innsky air fryer basket with avocado oil. Preheat the Innsky air fryer to 340°F. Place the parsley, chives, garlic, lemon juice, salt, sage, rosemary, fennel, and red pepper flakes in a food processor and puree until a smooth paste forms. Place the chicken breasts on a cutting board and rub the paste all over the tops. With a short end facing you, roll each breast up like a jelly roll to make a log and secure it with toothpicks. Wrap 2 slices of bacon around each chicken breast log to cover the entire breast. Secure the bacon with toothpicks.

2 Place the chicken breast logs in the Innsky air fryer basket and cook for 5 minutes, flip the logs over, and cook for another 5 minutes. Increase the heat to 390°F and cook until the bacon is crisp, about 5 minutes more. Remove the toothpicks and garnish with fresh rosemary sprigs, if desired, before serving.

Fried Chicken Sandwich

PREP: 10 MINUTES • COOK TIME: 15 MINUTES • TOTAL: 25 MINUTES • SERVES: 4

Ingredients

4 small skinless, boneless chicken thighs
3/4 cups low-fat buttermilk
2 teaspoons garlic powder
1/2 teaspoon salt
1/2 teaspoon pepper
1 cup all-purpose flour

4 potato rolls
Shredded romaine
Sliced tomatoes
Sliced pickles
Hot sauce

Instructions:

1. Place your chicken in a large bowl and mix in buttermilk, garlic powder, salt, and pepper. Place the flour in a separate bowl.
   Pull the chicken out of the buttermilk mix, letting excess drip off, cover in flour, dip in buttermilk mix, then cover in flour again.

2. Using your Air fryer, cook at 360 degrees for 15 minutes, flipping the chicken halfway through. Serve on a potato roll with lettuce, tomatoes, pickles, and hot sauce.

Chicken Paillard

PREP: 10 MINUTES • COOK TIME: 10 MINUTES • TOTAL: 20 MINUTES • SERVES: 2

Ingredients

2 large eggs, room temperature
1 tablespoon water
½ cup powdered Parmesan cheese
2 teaspoons dried thyme leaves
1 teaspoon ground black pepper
2 (5-ounce) boneless, skinless chicken breasts, pounded to ½ inch thick

LEMON BUTTER SAUCE:
2 tablespoons unsalted butter, melted
2 teaspoons lemon juice
¼ teaspoon finely chopped fresh thyme leaves, plus more for garnish
⅛ teaspoon fine sea salt
Lemon slices, for serving

Instructions:

1. Spray the Innsky air fryer basket with avocado oil. Preheat the Innsky air fryer to 390°F.
   Beat the eggs in a shallow dish, then add the water and stir well.
   In a separate shallow dish, mix together the Parmesan, thyme, and pepper until well combined.
   One at a time, dip the chicken breasts in the eggs and let any excess drip off, then dredge both sides of the chicken in the Parmesan mixture. As you finish, set the coated chicken in the air fryer basket.

2. Cook the chicken in the air fryer for 5 minutes, then flip the chicken and cook for another 5 minutes, or until cooked through and the internal temperature reaches 165°F.
   While the chicken cooks, make the lemon butter sauce: In a small bowl, mix together all the sauce ingredients until well combined.
   Plate the chicken and pour the sauce over it. Garnish with chopped fresh thyme and serve with lemon slices.
   Store leftovers in an airtight container in the refrigerator for up to 4 days. Reheat in a preheated 390°F air fryer for 5 minutes, or until heated through.

Per serving: Calories 526; Fat 33g; Protein 53g; Total carbs 3g; Fiber 1g

General Tso's Chicken
PREP: 10 MINUTES • COOK TIME: 20 MINUTES • TOTAL: 30 MINUTES • SERVES: 4

Ingredients

1 pound boneless, skinless chicken breasts or thighs, cut into 1-inch cubes

Fine sea salt and ground black pepper

GENERAL TSO'S SAUCE:

½ cup chicken broth

⅓ cup Swerve confectioners'-style sweetener

¼ cup coconut vinegar or unseasoned rice vinegar

¼ cup thinly sliced green onions, plus more for garnish if desired

1 tablespoon plus 1¼ teaspoons wheat-free tamari, or ¼ cup coconut aminos

3 small dried red chiles, chopped

1 clove garlic, minced

1½ teaspoons grated fresh ginger

1 teaspoon toasted sesame oil

¼ teaspoon guar gum (optional)

FOR SERVING (OPTIONAL):

Fried Cauliflower Rice

Sautéed broccoli rabe

FOR GARNISH (OPTIONAL):

Diced red chiles

Red pepper flakes

Sesame seeds

Instructions:

1   Preheat the Innsky air fryer to 400°F. Very lightly season the chicken on all sides with salt and pepper.

2   Place the chicken in a single layer in a pie pan that fits in the air fryer and cook for 5 minutes. While the chicken cooks, make the sauce: In a small bowl, stir together all the sauce ingredients except the guar gum until well combined. Sift in the guar gum (if using) and whisk until well combined. Pour the sauce over the chicken and, stirring every 5 minutes, cook for another 12 to 15 minutes, until the sauce is bubbly and thick and the chicken is cooked through and the internal temperature reaches 165°F. If you want the sauce to be even thicker and more flavorful, remove the chicken and return the sauce to the air fryer to cook for an additional 5 to 10 minutes. Transfer the chicken to a large bowl. Serve with fried cauliflower rice and sautéed broccoli rabe, if desired, and garnish with diced red chiles, sliced green onions, red pepper flakes, and sesame seeds, if desired. Store leftovers in an airtight container in the refrigerator for up to 4 days. Reheat in a preheated 375°F air fryer for 5 minutes, or until heated through.

Per serving: Calories 254; Fat 10g; Protein 34g; Total carbs 5g; Fiber 1g

Minty Chicken-Fried Pork Chops
PREP: 10 MINUTES • COOK TIME: 30 MINUTES • TOTAL: 40 MINUTES • SERVES: 6

Ingredients

4 medium-sized pork chops

1 cup of breadcrumbs

2 medium-sized eggs

Pinch of salt and pepper

½ tablespoon of mint, either dried and ground; or fresh, rinsed, and finely chopped

Instructions:

1.   Cover the basket of the Air fryer with a lining of tin foil, leaving the edges uncovered to allow air to circulate through the basket. Preheat the Innsky air fryer to 350 degrees. In a mixing bowl, beat the eggs until fluffy and until the yolks and whites are fully combined, and set aside. In a separate mixing bowl, combine the breadcrumbs, mint, salt and pepper, and set aside. One by one, dip each raw pork chop into the bowl with dry ingredients, coating all sides; then submerge into the bowl with wet ingredients, then dip again into the dry ingredients. Lay the coated pork chops on the foil covering the Air fryer basket, in a single flat layer.

2.   Set the Innsky air fryer timer for 15 minutes. After 15 minutes, the Air fryer will turn off and the pork should be mid-way cooked and the breaded coating starting to brown. Using tongs, turn each piece of steak over to ensure a full all-over fry. ReSet the Innsky air fryer to 320 degrees for 15 minutes. After 15 minutes remove the fried pork chops using tongs and set on a serving plate.

Crispy Southern Fried Chicken
PREP: 10 MINUTES • COOK TIME: 25 MINUTES • TOTAL: 35 MINUTES • SERVES: 4

Ingredients

1 tsp. cayenne pepper
2 tbsp. mustard powder
2 tbsp. oregano
2 tbsp. thyme
3 tbsp. coconut milk

1 beaten egg
¼ C. cauliflower
¼ C. gluten-free oats
8 chicken drumsticks

Instructions:

1   Ensure the Air fryer is preheated to 350 degrees.
    Lay out chicken and season with pepper and salt on all sides.
    Add all other ingredients to a blender, blending till a smooth-like breadcrumb mixture is created. Place in a bowl and add a beaten egg to another bowl.
    Dip chicken into breadcrumbs, then into egg, and breadcrumbs once more.
2   Place coated drumsticks into the Air fryer. Set temperature to 350°F, and set time to 20 minutes and cook 20 minutes. Bump up the temperature to 390 degrees and cook another 5 minutes till crispy.
    Per Serving: Calories: 504; Fat: 18g; Protein:35g; Sugar:5g

Chicken Strips with Satay Sauce
PREP: 5 MINUTES • COOK TIME: 10 MINUTES • TOTAL: 15 MINUTES • SERVES: 4

Ingredients

4 (6-ounce) boneless, skinless chicken breasts, sliced into 16 (1-inch) strips
1 teaspoon fine sea salt
1 teaspoon paprika
SAUCE:
¼ cup creamy almond butter
2 tablespoons chicken broth
1½ tablespoons coconut vinegar or unseasoned rice vinegar
1 clove garlic, minced
1 teaspoon peeled and minced fresh ginger

½ teaspoon hot sauce
⅛ teaspoon stevia glycerite, or 2 to 3 drops liquid stevia
FOR GARNISH/SERVING (OPTIONAL):
¼ cup chopped cilantro leaves
Red pepper flakes
Sea salt flakes
Thinly sliced red, orange, and yellow bell peppers
Special equipment:
16 wooden or bamboo skewers, soaked in water for 15 minutes

Instructions:

1. Spray the Innsky air fryer basket with avocado oil. Preheat the Innsky air fryer to 400°F.
    Thread the chicken strips onto the skewers. Season on all sides with the salt and paprika.
1. Place the chicken skewers in the Innsky air fryer basket and cook for 5 minutes, flip, and cook for another 5 minutes, until the chicken is cooked through and the internal temperature reaches 165°F.
    While the chicken skewers cook, make the sauce: In a medium-sized bowl, stir together all the sauce ingredients until well combined. Taste and adjust the sweetness and heat to your liking.
    Garnish the chicken with cilantro, red pepper flakes, and salt flakes, if desired, and serve with sliced bell peppers, if desired. Serve the sauce on the side.
    Store leftovers in an airtight container in the fridge for up to 4 days or in the freezer for up to a month. Reheat in a preheated 350°F air fryer for 3 minutes per side, or until heated through.
Per serving: Calories 359; Fat 16g; Protein 49g; Total carbs 2g, Fiber 1g

Bacon Lovers' Stuffed Chicken
PREP: 10 MINUTES • COOK TIME: 20 MINUTES • TOTAL: 30 MINUTES • SERVES: 4

Ingredients

4 (5-ounce) boneless, skinless chicken breasts, pounded to ¼ inch thick

2 packages Boursin cheese

8 slices thin-cut bacon or beef bacon

Sprig of fresh cilantro, for garnish

Instructions:

1. Spray the Innsky air fryer basket with avocado oil. Preheat the Innsky air fryer to 400°F. Place one of the chicken breasts on a cutting board. With a sharp knife held parallel to the cutting board, make a 1-inch-wide incision at the top of the breast. Carefully cut into the breast to form a large pocket, leaving a ½-inch border along the sides and bottom. Repeat with the other 3 chicken breasts. Snip the corner of a large resealable plastic bag to form a ¾-inch hole. Place the Boursin cheese in the bag and pipe the cheese into the pockets in the chicken breasts, dividing the cheese evenly among them. Wrap 2 slices of bacon around each chicken breast and secure the ends with toothpicks.

2. Place the bacon-wrapped chicken in the Innsky air fryer basket and cook until the bacon is crisp and the chicken's internal temperature reaches 165°F, about 18 to 20 minutes, flipping after 10 minutes. Garnish with a sprig of cilantro before serving, if desired. Store leftovers in an airtight container in the refrigerator for up to 4 days. Reheat in a preheated 400°F air fryer for 5 minutes, or until warmed through.

Chicken Roast with Pineapple Salsa
PREP: 10 MINUTES • COOK TIME: 45 MINUTES • TOTAL: 55 MINUTES • SERVES: 2

Ingredients

¼ cup extra virgin olive oil

¼ cup freshly chopped cilantro

1 avocado, diced

1-pound boneless chicken breasts

2 cups canned pineapples

2 teaspoons honey

Juice from 1 lime

Salt and pepper to taste

Instructions:

1. Preheat the Innsky air fryer to 390°F. Place the grill pan accessory in the air fryer. Season the chicken breasts with lime juice, olive oil, honey, salt, and pepper.

2. Place on the grill pan and cook for 45 minutes. Flip the chicken every 10 minutes to grill all sides evenly. Once the chicken is cooked, serve with pineapples, cilantro, and avocado.

Tex-Mex Turkey Burgers
PREP: 10 MINUTES • COOK TIME: 15 MINUTES • TOTAL: 25 MINUTES • SERVES: 4

Ingredients

⅓ cup finely crushed corn tortilla chips

1 egg, beaten

¼ cup salsa

⅓ cup shredded pepper Jack cheese

Pinch salt

Freshly ground black pepper

1 pound ground turkey

1 tablespoon olive oil

1 teaspoon paprika

Instructions:

1. In a medium bowl, combine the tortilla chips, egg, salsa, cheese, salt, and pepper, and mix well. Add the turkey and mix gently but thoroughly with clean hands. Form the meat mixture into patties about ½ inch thick. Make an indentation in the center of each patty with your thumb so the burgers don't puff up while cooking. Brush the patties on both sides with the olive oil and sprinkle with paprika.

2. Put in the Air fryer basket. Grill for 14 to 16 minutes or until the meat registers at least 165°F.

Chicken Pesto Parmigiana

PREP: 10 MINUTES • COOK TIME: 23 MINUTES • TOTAL: 33 MINUTES • SERVES: 4

Ingredients

2 large eggs
1 tablespoon water
Fine sea salt and ground black pepper
1 cup powdered Parmesan cheese
2 teaspoons Italian seasoning

4 (5-ounce) boneless, skinless chicken breasts or thighs, pounded to ¼ inch thick
1 cup pesto
1 cup shredded mozzarella cheese
Finely chopped fresh basil, for garnish
Grape tomatoes, halved, for serving

Instructions:

1. Spray the Innsky air fryer basket with avocado oil. Preheat the Innsky air fryer to 400°F.Crack the eggs into a shallow baking dish, add the water and a pinch each of salt and pepper, and whisk to combine. In another shallow baking dish, stir together the Parmesan and Italian seasoning until well combined. Season the chicken breasts well on both sides with salt and pepper. Dip one chicken breast in the eggs and let any excess drip off, then dredge both sides of the breast in the Parmesan mixture. Spray the breast with avocado oil and place it in the air fryer basket. Repeat with the remaining 3 chicken breasts.

2.

   Cook the chicken in the air fryer for 20 minutes, or until the internal temperature reaches 165°F and the breading is golden brown, flipping halfway through.

   Dollop each chicken breast with ¼ cup of the pesto and top with the mozzarella. Return the breasts to the air fryer and cook for 3 minutes, or until the cheese is melted. Garnish with basil and serve with halved grape tomatoes on the side, if desired.

Crispy Taco Chicken

PREP: 10 MINUTES • COOK TIME: 23 MINUTES • TOTAL: 33 MINUTES • SERVES: 4

Ingredients

2 large eggs
1 tablespoon water
Fine sea salt and ground black pepper
1 cup pork dust
1 teaspoon ground cumin
1 teaspoon smoked paprika

4boneless, skinless chicken breasts or thighs, pounded to ¼ inch thick
1 cup salsa
1 cup shredded Monterey Jack cheese
Sprig of fresh cilantro, for garnish

Instructions:

1. Spray the Innsky air fryer basket with avocado oil. Preheat the Innsky air fryer to 400°F. Crack the eggs into a shallow baking dish, add the water and a pinch each of salt and pepper, and whisk to combine. In another shallow baking dish, stir together the pork dust, cumin, and paprika until well combined. Season the chicken breasts well on both sides with salt and pepper. Dip 1 chicken breast in the eggs and let any excess drip off, then dredge both sides of the chicken breast in the pork dust mixture. Spray the breast with avocado oil and place it in the air fryer basket. Repeat with the remaining 3 chicken breasts.

2. Cook the chicken in the air fryer for 20 minutes, or until the internal temperature reaches 165°F and the breading is golden brown, flipping halfway through.

   Dollop each chicken breast with ¼ cup of the salsa and top with ¼ cup of the cheese. Return the breasts to the air fryer and cook for 3 minutes, or until the cheese is melted. Garnish with cilantro before serving, if desired.

Bacon-Wrapped Stuffed Chicken Breasts

PREP: 15 MINUTES • COOK TIME: 30 MINUTES • TOTAL: 45 MINUTES • SERVES: 4

Ingredients

½ cup chopped frozen spinach, thawed and squeezed dry

¼ cup cream cheese, softened

¼ cup grated Parmesan cheese

1 jalapeño, seeded and chopped

½ teaspoon kosher salt

1 teaspoon black pepper

2 large boneless, skinless chicken breasts, butterflied and pounded to ½-inch thickness

4 teaspoons salt-free Cajun seasoning

6 slices bacon

Instructions:

1. In a small bowl, combine the spinach, cream cheese, Parmesan cheese, jalapeño, salt, and pepper. Stir until well combined. Place the butterflied chicken breasts on a flat surface. Spread the cream cheese mixture evenly across each piece of chicken. Starting with the narrow end, roll up each chicken breast, ensuring the filling stays inside. Season chicken with the Cajun seasoning, patting it in to ensure it sticks to the meat. Wrap each breast in 3 slices of bacon.

2. Place in the air fryer basket. Set the Innsky air fryer to 350°F for 30 minutes. Use a meat thermometer to ensure the chicken has reached an internal temperature of 165°F.
   Let the chicken stand 5 minutes before slicing each rolled-up breast in half to serve.

Thai Tacos with Peanut Sauce

PREP: 10 MINUTES • COOK TIME: 6 MINUTES • TOTAL: 16 MINUTES • SERVES: 4

Ingredients

1 pound ground chicken

¼ cup diced onions (about 1 small onion)

2 cloves garlic, minced

¼ teaspoon fine sea salt

SAUCE:

¼ cup creamy peanut butter, room temperature

2 tablespoons chicken broth, plus more if needed

2 tablespoons lime juice

2 tablespoons grated fresh ginger

2 tablespoons wheat-free tamari or coconut aminos

1½ teaspoons hot sauce

5 drops liquid stevia (optional)

FOR SERVING:

2 small heads butter lettuce, leaves separated

Lime slices (optional)

FOR GARNISH (OPTIONAL):

Cilantro leaves

Shredded purple cabbage

Sliced green onions

Instructions:

1. Preheat the Innsky air fryer to 350°F. Place the ground chicken, onions, garlic, and salt in a 6-inch pie pan or a dish that will fit in your air fryer. Break up the chicken with a spatula.

2. Place in the air fryer and cook for 5 minutes, or until the chicken is browned and cooked through. Break up the chicken again into small crumbles. Make the sauce: In a medium-sized bowl, stir together the peanut butter, broth, lime juice, ginger, tamari, hot sauce, and stevia (if using) until well combined. If the sauce is too thick, add another tablespoon or two of broth. Taste and add more hot sauce if desired. Add half of the sauce to the pan with the chicken. Cook for another minute, until heated through, and stir well to combine.
   Assemble the tacos: Place several lettuce leaves on a serving plate. Place a few tablespoons of the chicken mixture in each lettuce leaf and garnish with cilantro leaves, purple cabbage, and sliced green onions, if desired. Serve the remaining sauce on the side. Serve with lime slices, if desired.
   Store leftover meat mixture in an airtight container in the refrigerator for up to 4 days; store leftover sauce, lettuce leaves, and garnishes separately. Reheat the meat mixture in a lightly greased pie pan in a preheated 350°F air fryer for 3 minutes, or until heated through.

Per serving: Calories 350; Fat 17g; Protein 39g; Total carbs 11g; Fiber 3g

Chicken Cordon Bleu

PREP: 15 MINUTES • COOK TIME: 25 MINUTES • TOTAL: 15 MINUTES • SERVES: 4

Ingredients

2 (8-ounce) boneless, skinless chicken breasts
8 thin slices deli ham
4 slices Swiss cheese
½ teaspoon black pepper

1 large egg
¾ cup panko bread crumbs or crushed pork rinds
Vegetable oil spray

Instructions:

1. Cut the chicken breasts horizontally in half to create four thin chicken cutlets. Place two slices of ham and one slice of cheese on each piece of chicken. Sprinkle with the pepper. Starting on a short end, roll up each piece of chicken and secure with a toothpick. In a shallow bowl, beat the egg. Place the bread crumbs on a plate. Dip each chicken roll in the egg, then roll in the bread crumbs to coat. Spray all sides generously with vegetable oil spray.
2. Place the chicken in the air fryer basket. Set the Innsky air fryer to 350°F for 25 minutes, turning the chicken and spraying with oil spray halfway through the cooking time. Use a meat thermometer to ensure the chicken has reached an internal temperature of 165°F.

Brazilian Tempero Baiano Chicken Drumsticks

PREP: 5 MINUTES PLUS 30 MINUTES TO MARINATE • COOK TIME: 20 MINUTES • TOTAL: 55 MINUTES SERVES: 4

Ingredients

1 teaspoon cumin seeds
1 teaspoon dried oregano
1 teaspoon dried parsley
1 teaspoon ground turmeric
½ teaspoon coriander seeds
1 teaspoon kosher salt

½ teaspoon black peppercorns
½ teaspoon cayenne pepper
¼ cup fresh lime juice
2 tablespoons olive oil
1½ pounds chicken drumsticks

Instructions:

1. In a clean coffee grinder or spice mill, combine the cumin, oregano, parsley, turmeric, coriander seeds, salt, peppercorns, and cayenne. Process until finely ground. In a small bowl, combine the ground spices with the lime juice and oil. Place the chicken in a resealable plastic bag. Add the marinade, seal, and massage until the chicken is well coated. Marinate at room temperature for 30 minutes or in the refrigerator for up to 24 hours.
2. When you are ready to cook, place the drumsticks skin side up in the air fryer basket. Set the Innsky air fryer to 400°F for 20 to 25 minutes, turning the legs halfway through the cooking time. Use a meat thermometer to ensure that the chicken has reached an internal temperature of 165°F. Serve with plenty of napkins.

Air Fryer Turkey Breast

PREP: 5 MINUTES • COOK TIME: 60 MINUTES • TOTAL: 65 MINUTES • SERVES: 6

Ingredients

Pepper and salt
1 oven-ready turkey breast

Turkey seasonings of choice

Instructions:

1. Preheat the Innsky air fryer to 350 degrees.
   Season turkey with pepper, salt, and other desired seasonings.
   Place turkey in air fryer basket.
2. Set temperature to 350°F, and set time to 60 minutes. Cook 60 minutes. The meat should be at 165 degrees when done. Allow to rest 10-15 minutes before slicing. Enjoy.

Per Serving: Calories: 212; Fat: 12g; Protein:24g; Sugar:0g

Cheese Stuffed Chicken
PREP: 5 MINUTES • COOK TIME: 30 MINUTES • TOTAL: 35 MINUTES • SERVES: 4

Ingredients

1 tablespoon creole seasoning
1 tablespoon olive oil
1 teaspoon garlic powder
1 teaspoon onion powder

4 chicken breasts, butterflied and pounded
4 slices Colby cheese
4 slices pepper jack cheese

Instructions:

1   Preheat the Innsky air fryer to 390°F. Place the grill pan accessory in the air fryer.
   Create the dry rub by mixing in a bowl the creole seasoning, garlic powder, and onion powder. Season with salt and pepper if desired. Rub the seasoning on to the chicken. Place the chicken on a working surface and place a slice each of pepper jack and Colby cheese. Fold the chicken and secure the edges with toothpicks. Brush chicken with olive oil.
2   Grill for 30 minutes and make sure to flip the meat every 10 minutes.
Per Serving: Calories: 27; Fat: 45.9g; Protein:73.1g; Sugar:0g

Chicken Jalfrezi
PREP: 15 MINUTES • COOK TIME: 15 MINUTES • TOTAL: 30 MINUTES • SERVES: 4

Ingredients

For the Chicken

  1pound boneless, skinless chicken thighs, cut into
     2 or 3 pieces each
  1 medium onion, chopped
  1 large green bell pepper, stemmed, seeded, and
     chopped
  2 tablespoons olive oil
  1 teaspoon ground turmeric
  1 teaspoon Garam Masala

1 teaspoon kosher salt
½ to 1 teaspoon cayenne pepper
For the Sauce
¼ cup tomato sauce
1 tablespoon water
1 teaspoon Garam Masala
½ teaspoon kosher salt
½ teaspoon cayenne pepper
Side salad, rice, or naan bread, for serving

Instructions:

1.   For the chicken: In a large bowl, combine the chicken, onion, bell pepper, oil, turmeric, garam masala, salt, and cayenne. Stir and toss until well combined.
2.   Place the chicken and vegetables in the air fryer basket. Set the Innsky air fryer to 350°F for 15 minutes, stirring and tossing halfway through the cooking time. Use a meat thermometer to ensure the chicken has reached an internal temperature of 165°F. Meanwhile, for the sauce: In a small microwave-safe bowl, combine the tomato sauce, water, garam masala, salt, and cayenne. Microwave on high for 1 minute. Remove and stir. Microwave for another minute; set aside. When the chicken is cooked, remove and place chicken and vegetables in a large bowl. Pour the sauce over all. Stir and toss to coat the chicken and vegetables evenly. Serve with rice, naan, or a side salad.

Mustard Chicken Tenders
PREP: 5 MINUTES • COOK TIME: 20 MINUTES • TOTAL: 25 MINUTES • SERVES: 4

Ingredients

½ C. coconut flour
1 tbsp. spicy brown mustard

2 beaten eggs
1 pound of chicken tenders

Instructions:

1   Season tenders with pepper and salt.
   Place a thin layer of mustard onto tenders and then dredge in flour and dip in egg.
2   Add to the Air fryer, set temperature to 390°F, and set time to 20 minutes.

Orange Curried Chicken Stir-Fry
PREP: 10 MINUTES • COOK TIME: 18 MINUTES • TOTAL: 28 MINUTES • SERVES: 4

Ingredients
¾ pound boneless, skinless chicken thighs, cut into 1-inch pieces
1 yellow bell pepper, cut into 1½-inch pieces
1 small red onion, sliced
Olive oil for misting

¼ cup chicken stock
2 tablespoons honey
¼ cup orange juice
1 tablespoon cornstarch
2 to 3 teaspoons curry powder

Instructions:

1. Put the chicken thighs, pepper, and red onion in the Innsky air fryer basket and mist with olive oil.
2. Cook for 12 to 14 minutes or until the chicken is cooked to 165°F, shaking the basket halfway through cooking time. Remove the chicken and vegetables from the Innsky air fryer basket and set aside. In a 6-inch metal bowl, combine the stock, honey, orange juice, cornstarch, and curry powder, and mix well. Add the chicken and vegetables, stir, and put the bowl in the basket. Return the basket to the Air fryer and cook for 2 minutes. Remove and stir, then cook for 2 to 3 minutes or until the sauce is thickened and bubbly.

Curry Mustard Chicken
PREP: 10 MINUTES • COOK TIME: 15 MINUTES • TOTAL: 25 MINUTES • SERVES: 4

Ingredients
6 tablespoons mayonnaise
2 tablespoons coarse-ground mustard
2 teaspoons honey (optional)
2 teaspoons curry powder

1 teaspoon kosher salt
1 teaspoon cayenne pepper
1 pound chicken tenders

Instructions:

1. In a large bowl, whisk together the mayonnaise, mustard, honey (if using), curry powder, salt, and cayenne. Transfer half of the mixture to a serving bowl to serve as a dipping sauce. Add the chicken tenders to the large bowl and toss and stir until well coated.
2. Place the tenders in the air fryer basket. Set the Innsky air fryer to 350°F for 15 minutes. Use a meat thermometer to ensure the chicken has reached an internal temperature of 165°F.
Serve the chicken with the dipping sauce.

Chicken BBQ with Sweet And Sour Sauce
PREP: 5 MINUTES • COOK TIME: 40 MINUTES • TOTAL: 45 MINUTES • SERVES: 6

Ingredients
¼ cup minced garlic
¼ cup tomato paste
¾ cup minced onion
¾ cup sugar
1 cup soy sauce

1 cup water
1 cup white vinegar
6 chicken drumsticks
Salt and pepper to taste

Instructions:

1. Place all Ingredients in a Ziploc bag. Allow to marinate for at least 2 hours in the fridge. Preheat the Innsky air fryer to 390°F. Place the grill pan accessory in the air fryer.
2. Grill the chicken for 40 minutes. Flip the chicken every 10 minutes for even grilling. Meanwhile, pour the marinade in a saucepan and heat over medium flame until the sauce thickens. Before serving the chicken, brush with the glaze.

Cilantro Chicken Kebabs

PREP: 20 MINUTES PLUS 15 MINUTES TO MARINATE • COOK TIME: 10 MINUTES • TOTAL: 45 MINUTES •SERVES: 4

Ingredients

For the Chutney
½ cup unsweetened shredded coconut
½ cup hot water
2 cups fresh cilantro leaves, roughly chopped
¼ cup fresh mint leaves, roughly chopped
6 cloves garlic, roughly chopped
1 jalapeño, seeded and roughly chopped

¼ to ¾ cup water, as needed
Juice of 1 lemon
For the Chicken
1pound boneless, skinless chicken thighs, cut crosswise into thirds
Olive oil spray

Instructions:

1. For the chutney: In a blender or food processor, combine the coconut and hot water; set aside to soak for 5 minutes. To the processor, add the cilantro, mint, garlic, and jalapeño, along with ¼ cup water. Blend at low speed, stopping occasionally to scrape down the sides. Add the lemon juice. With the blender or processor running, add only enough additional water to keep the contents moving. Turn the blender to high once the contents are moving freely and blend until the mixture is puréed.
    For the chicken: Place the chicken pieces in a large bowl. Add ¼ cup of the chutney and mix well to coat. Set aside the remaining chutney to use as a dip. Marinate the chicken for 15 minutes at room temperature.

2.
    Spray the Innsky air fryer basket with olive oil spray. Arrange the chicken in the air fryer basket. Set the Innsky air fryer to 350°F for 10 minutes. Use a meat thermometer to ensure that the chicken has reached an internal temperature of 165°F. Serve the chicken with the remaining chutney.

Chicken Pot Pie with Coconut Milk

PREP: 5 MINUTES • COOK TIME: 30 MINUTES • TOTAL: 40 MINUTES • SERVES: 8

Ingredients

¼ small onion, chopped
½ cup broccoli, chopped
¾ cup coconut milk
1 cup chicken broth
1/3 cup coconut flour
1-pound ground chicken

2 cloves of garlic, minced
2 tablespoons butter
4 ½ tablespoons butter, melted
4 eggs
Salt and pepper to taste

Instructions:

1   Preheat the Innsky air fryer for 5 minutes.
    Place 2 tablespoons butter, broccoli, onion, garlic, coconut milk, chicken broth, and ground chicken in a baking dish that will fit in the air fryer. Season with salt and pepper to taste.
    In a mixing bowl, combine the butter, coconut flour, and eggs.
    Sprinkle evenly the top of the chicken and broccoli mixture with the coconut flour dough.
    Place the dish in the air fryer.

2   Cook for 30 minutes at 325°F.

Per Serving: Calories: 366; Fat: 29.5g; Protein:21.8g; Sugar:4g

Chicken Nuggets

PREP: 10 MINUTES • COOK TIME: 20 MINUTES • TOTAL: 30 MINUTES • SERVES: 4

Ingredients

1pound boneless, skinless chicken breasts
Chicken seasoning or rub
Salt
Pepper

2 eggs
6 tablespoons bread crumbs
2 tablespoons panko bread crumbs
Cooking oil

Instructions:

1. Cut the chicken breasts into 1-inch pieces.

    In a large bowl, combine the chicken pieces with chicken seasoning, salt, and pepper to taste. In a small bowl, beat the eggs. In another bowl, combine the bread crumbs and panko. Dip the chicken pieces in the eggs and then the bread crumbs.

    Place the nuggets in the Air fryer. Do not overcrowd the basket. Cook in batches. Spray the nuggets with cooking oil.

2.

    Cook for 4 minutes. Open the Air fryer and shake the basket. Cook for an additional 4 minutes. Remove the cooked nuggets from the Air fryer, then repeat steps 5 and 6 for the remaining chicken nuggets. Cool before serving.

Per Serving: Calories: 206; Fat: 5g; Protein:31g; Fiber:1g

Cheesy Chicken Fritters
PREP: 5 MINUTES • COOK TIME: 20 MINUTES • TOTAL: 25 MINUTES • SERVES: 17 FRITTERS

Ingredients

Chicken Fritters:
½ tsp. salt
1/8 tsp. pepper
1 ½ tbsp. fresh dill
1 1/3 C. shredded mozzarella cheese
1/3 C. coconut flour
1/3 C. vegan mayo
2 eggs

1 ½ pounds chicken breasts
Garlic Dip:
1/8 tsp. pepper
¼ tsp. salt
½ tbsp. lemon juice
1 pressed garlic cloves
1/3 C. vegan mayo

Instructions:

1. Slice chicken breasts into 1/3" pieces and place in a bowl. Add all remaining fritter ingredients to the bowl and stir well. Cover and chill 2 hours or overnight. Ensure your air fryer is preheated to 350 degrees. Spray basket with a bit of olive oil.

2. Add marinated chicken to air fryer. Set temperature to 350°F, and set time to 20 minutes and cook 20 minutes, making sure to turn halfway through cooking process. To make the dipping sauce, combine all the dip ingredients until smooth.

French Garlic Chicken
PREP: 10 MINUTES PLUS 30 MINUTES TO MARINATE • COOK TIME: 27 MINUTES • TOTAL: 67 MINUTES • SERVES: 4

Ingredients

2 tablespoon extra-virgin olive oil
1 tablespoon Dijon mustard
1 tablespoon apple cider vinegar
3 cloves garlic, minced
2 teaspoons herbes de Provence
½ teaspoon kosher salt

1 teaspoon black pepper
1 pound boneless, skinless chicken thighs, halved crosswise
2 tablespoons butter
8 cloves garlic, chopped
¼ cup heavy whipping cream

Instructions:

1. In a small bowl, combine the olive oil, mustard, vinegar, minced garlic, herbes de Provence, salt, and pepper. Use a wire whisk to emulsify the mixture. Pierce the chicken all over with a fork to allow the marinade to penetrate better. Place the chicken in a resealable plastic bag, pour the marinade over, and seal. Massage until the chicken is well coated. Marinate at room temperature for 30 minutes or in the refrigerator for up to 24 hours.

2. When you are ready to cook, place the butter and chopped garlic in a 7 × 3-inch round heatproof pan and place it in the air fryer basket. Set the Innsky air fryer to 400°F for 5 minutes, or until the butter has melted and the garlic is sizzling. Add the chicken and the marinade to the seasoned butter. Set the Innsky air fryer to 350°F for 15 minutes. Use a meat thermometer to ensure the chicken has reached an internal temperature of 165°F. Transfer the chicken to a plate and cover lightly with foil to keep warm. Add the cream to the pan, stirring to combine with the garlic, butter, and cooking juices. Place the pan in the air fryer basket. Set the Innsky air fryer to 350°F for 7 minutes. Pour the thickened sauce over the chicken and serve.

Crusted Chicken Tenders

PREP: 5 MINUTES • COOK TIME: 15 MINUTES • TOTAL: 20 MINUTES • SERVES: 3

Ingredients
½ cup all-purpose flour
2 eggs, beaten
½ cup seasoned breadcrumbs

Salt and freshly ground black pepper, to taste
2 tablespoons olive oil
¾ pound chicken tenders

Instructions:
1. In a bowl, place the flour. In a second bowl, place the eggs. In a third bowl, mix together breadcrumbs, salt, black pepper and oil. Coat the chicken tenders in the flour, Then dip into the eggs and finally coat with the breadcrumbs mixture evenly.
2. Preheat the Innsky air fryer to 330 degrees F. Arrange the chicken tenderloins in Air fryer basket. Cook for about 10 minutes. Now, Set the Innsky air fryer to 390 degrees F. Cook for about 5 minutes further.

Ginger Chicken

PREP: 10 MINUTES PLUS 30 MINUTES TO MARINATE • COOK TIME: 10 MINUTES • TOTAL: 50 MINUTES • SERVES: 4

Ingredients
¼ cup julienned peeled fresh ginger
2 tablespoons vegetable oil
1 tablespoon honey
1 tablespoon soy sauce
1 tablespoon ketchup
1 teaspoon Garam Masala
1 teaspoon ground turmeric

¼ teaspoon kosher salt
½ teaspoon cayenne pepper
Vegetable oil spray
1pound boneless, skinless chicken thighs, cut crosswise into thirds
¼ cup chopped fresh cilantro, for garnish

Instructions:
1. In a small bowl, combine the ginger, oil, honey, soy sauce, ketchup, garam masala, turmeric, salt, and cayenne. Whisk until well combined. Place the chicken in a resealable plastic bag and pour the marinade over. Seal the bag and massage to cover all of the chicken with the marinade. Marinate at room temperature for 30 minutes or in the refrigerator for up to 24 hours.
2. Spray the Innsky air fryer basket with vegetable oil spray and add the chicken and as much of the marinade and julienned ginger as possible. Set the Innsky air fryer to 350°F for 10 minutes. Use a meat thermometer to ensure the chicken has reached an internal temperature of 165°F. To serve, garnish with cilantro.

Chicken BBQ Recipe from Peru

PREP: 5 MINUTES • COOK TIME: 40 MINUTES • TOTAL: 45 MINUTES • SERVES: 4

Ingredients
½ teaspoon dried oregano
1 teaspoon paprika
1/3 cup soy sauce
2 ½ pounds chicken, quartered

2 tablespoons fresh lime juice
2 teaspoons ground cumin
5 cloves of garlic, minced

Instructions:
1. Place all Ingredients in a Ziploc bag and shake to mix everything.
   Allow to marinate for at least 2 hours in the fridge.
   Preheat the Innsky air fryer to 390°F. Place the grill pan accessory in the air fryer.
2. Grill the chicken for 40 minutes making sure to flip the chicken every 10 minutes for even grilling.

Per Serving: Calories: 377; Fat: 11.8g; Protein:59.7g; Sugar:0g

Thai Basll Chicken

PREP: 5 MINUTES • COOK TIME: 20 MINUTES • TOTAL: 25 MINUTES • SERVES: 4

Ingredients

4 Chicken Breasts
1 Onion
2 Bell Peppers
2 Hot Peppers
1 Tbsp Olive Oil
3 Tbsps Fish Sauce
2 Tbsps Oyster Sauce

3 Tbsps Sweet Chili Sauce
1 Tbsp Soy Sauce
1 Quart Chicken Broth
1 Tbsp Garlic Powder
1 Tbsp Chili Powder
1 Cup Thai Basil

Instructions:

1 Wash the breasts and boil them in the chicken broth for 10 minutes, then lower to simmer for another 10 minutes until tender. Take them out of the broth and allow to cool. Using two forks, tear the chicken into shreds. Toss the shreds with the garlic powder, chili powder, and salt and pepper to taste

2 Preheat the Innsky air fryer to 390 degrees and cook the chicken shreds for 20 minutes, at which point they will get dark brown and crispy. They will soften up as they absorb the juices from cooking with the veggies. While the chicken is cooking, cut the onions and peppers into thin slices. Add the olive oil to a wok and heat for a minute on medium high heat. Toss in all the veggies and sauté for 5 minutes. Add in the fish sauce, oyster sauce, soy sauce, sweet chili sauce, and stir well for 1 minute. Add the chicken and basil leaves and stir until the leaves have wilted. Serve over jasmine rice.

Air Fryer Chicken Parmesan

PREP: 5 MINUTES • COOK TIME: 9 MINUTES • TOTAL: 20 MINUTES • SERVES: 4

Ingredients

½ C. keto marinara
6 tbsp. mozzarella cheese
1 tbsp. melted ghee

2 tbsp. grated parmesan cheese
6 tbsp. gluten-free seasoned breadcrumbs
8-ounce chicken breasts

Instructions:

1 Ensure air fryer is preheated to 360 degrees. Spray the basket with olive oil. Mix parmesan cheese and breadcrumbs together. Melt ghee.
Brush melted ghee onto the chicken and dip into breadcrumb mixture.
Place coated chicken in the air fryer and top with olive oil.

2 Set temperature to 360°F, and set time to 6 minutes. Cook 2 breasts for 6 minutes and top each breast with a tablespoon of sauce and 1½ tablespoons of mozzarella cheese. Cook another 3 minutes to melt cheese.
Keep cooked pieces warm as you repeat the process with remaining breasts.

Per Serving: Calories: 251; Fat: 10g; Protein:31g; Sugar:0g

Spicy Chicken-Fried Steak with Peppercorn Gravy
PREP: 20 MINUTES • COOK TIME: 8 MINUTES • TOTAL: 28 MINUTES • SERVES: 2

Ingredients

For the Steaks
2 beef cube steaks (5 to 6 ounces each)
¾ cup all-purpose flour
2 teaspoons crumbled dried sage
½ teaspoon smoked paprika
½ teaspoon onion powder
½ teaspoon garlic powder
¼ teaspoon cayenne pepper
Kosher salt and black pepper
¾ cup buttermilk
1 large egg

1 teaspoon hot pepper sauce
Vegetable oil spray
For the Gravy
2 tablespoons butter
2 tablespoons all-purpose flour
¼ teaspoon garlic salt
½ teaspoon kosher salt
½ teaspoon cracked black pepper
1 cup whole milk
½ cup heavy whipping cream

Instructions:

1. For the steaks: Cut the steaks in half if needed to fit in the air fryer basket; set aside. In a shallow bowl, whisk the flour, sage, paprika, onion powder, garlic powder, cayenne, 1 teaspoon salt, and 1 teaspoon pepper. In a separate shallow bowl, whisk together the buttermilk, egg, and hot pepper sauce until combined. Using a paper towel, pat the steaks dry. Season to taste with salt and pepper. Allow to stand for 5 minutes, then pat dry again. Dredge the steaks in the flour mixture, shaking off any excess; then dip in the buttermilk mixture, allowing excess to drip off. Dredge once more in the flour mixture, shaking off excess. Place the breaded steaks on a baking sheet and press any of the remaining flour mixture onto the steaks, making sure that each steak is completely coated. Let stand for 10 minutes.

2. Place the steaks in the air fryer basket. Lightly coat with vegetable oil spray. Set the Innsky air fryer to 400°F for 8 minutes, carefully turning the steaks halfway through the cooking time and coating the other side with the oil spray. Use a meat thermometer to ensure the steaks have reached an internal temperature of 145°F. *Meanwhile, for the gravy:* In a small saucepan, melt the butter over low heat. Add the flour, garlic salt, kosher salt, and cracked pepper and whisk until smooth. Slowly add the milk and cream while continuing to whisk. Turn the heat to medium and cook, whisking occasionally, until the gravy thickens. Serve the steaks topped with the gravy.

Ricotta and Parsley Stuffed Turkey Breasts
PREP: 5 MINUTES • COOK TIME: 25 MINUTES • TOTAL: 30 MINUTES • SERVES: 4

Ingredients

1 turkey breast, quartered
1 cup Ricotta cheese
1/4 cup fresh Italian parsley, chopped
1 teaspoon garlic powder
1/2 teaspoon cumin powder

1 egg, beaten
1 teaspoon paprika
Salt and ground black pepper, to taste
Crushed tortilla chips
1 ½ tablespoons extra-virgin olive oil

Instructions:

1. Firstly, flatten out each piece of turkey breast with a rolling pin. Prepare three mixing bowls. In a shallow bowl, combine Ricotta cheese with the parsley, garlic powder, and cumin powder. Place the Ricotta/parsley mixture in the middle of each piece. Repeat with the remaining pieces of the turkey breast and roll them up. In another shallow bowl, whisk the egg together with paprika. In the third shallow bowl, combine the salt, pepper, and crushed tortilla chips. Dip each roll in the whisked egg, then, roll them over the tortilla chips mixture. Transfer prepared rolls to the Air fryer basket. Drizzle olive oil over all.

2. Cook at 350 degrees F for 25 minutes, working in batches. Serve warm, garnished with some extra parsley, if desired.

Cheesy Turkey-Rice with Broccoli
PREP: 5 MINUTES • COOK TIME: 40 MINUTES • TOTAL: 45 MINUTES • SERVES: 4

Ingredients
1 cup cooked, chopped turkey meat
1 tablespoon and 1-1/2 teaspoons butter, melted
1/2 (10 ounce) package frozen broccoli, thawed

1/2 (7 ounce) package whole wheat crackers, crushed
1/2 cup shredded Cheddar cheese
1/2 cup uncooked white rice

Instructions:

1   Bring to a boil 2 cups of water in a saucepan. Stir in rice and simmer for 20 minutes. Turn off fire and set aside.

Lightly grease baking pan of air fryer with cooking spray. Mix in cooked rice, cheese, broccoli, and turkey. Toss well to mix.

Mix well melted butter and crushed crackers in a small bowl. Evenly spread on top of rice.

2   For 20 minutes, cook on 360°F until tops are lightly browned.

Serve and enjoy.

Per Serving: Calories: 269; Fat: 11.8g; Protein:17g; Sugar:0g

Jerk Chicken Wings
PREP: 10 MINUTES • COOK TIME: 16 MINUTES • TOTAL: 26 MINUTES • SERVES: 6

Ingredients
1 tsp. salt
½ C. red wine vinegar
5 tbsp. lime juice
4 chopped scallions
1 tbsp. grated ginger
2 tbsp. brown sugar
1 tbsp. chopped thyme
1 tsp. white pepper
1 tsp. cayenne pepper

1 tsp. cinnamon
1 tbsp. allspice
1 Habanero pepper (seeds/ribs removed and chopped finely)
6 chopped garlic cloves
2 tbsp. low-sodium soy sauce
2 tbsp. olive oil
4 pounds of chicken wings

Instructions:

1   Combine all ingredients except wings in a bowl. Pour into a gallon bag and add chicken wings. Chill 2-24 hours to marinate.

Ensure your air fryer is preheated to 390 degrees.

Place chicken wings into a strainer to drain excess liquids.

2   Pour half of the wings into your Air fryer. Set temperature to 390°F, and set time to 16 minutes and cook 14-16 minutes, making sure to shake halfway through the cooking process.

Remove and repeat the process with remaining wings.

Per Serving: Calories: 374; Fat: 14g; Protein:33g; Sugar:4g

Pork Recipes

Pork Taquitos

PREP: 10 MINUTES • COOK TIME: 16 MINUTES • TOTAL: 26 MINUTES • SERVES: 8

Ingredients

1 juiced lime

10 whole wheat tortillas

2 ½ C. shredded mozzarella cheese

30 ounces of cooked and shredded pork tenderloin

Instructions:

1. Ensure your air fryer is preheated to 380 degrees.
   Drizzle pork with lime juice and gently mix.
   Heat up tortillas in the microwave with a dampened paper towel to soften.
   Add about 3 ounces of pork and ¼ cup of shredded cheese to each tortilla. Tightly roll them up.
   Spray the Innsky air fryer basket with a bit of olive oil.
2. Set temperature to 380°F, and set time to 10 minutes. Air fry taquitos 7-10 minutes till tortillas turn a slight golden color, making sure to flip halfway through cooking process.

Per Serving: Calories: 309; Fat: 11g; Protein:21g; Sugar:2g

Cajun Bacon Pork Loin Fillet

PREP: 10 MINUTES PLUS 1 HOUR TO MARINATE•COOK TIME: 20 MINUTES•TOTAL:1 HOUR 30MINUTES•SERVES: 6

Ingredients

1½ pounds pork loin fillet or pork tenderloin

3 tablespoons olive oil

2 tablespoons Cajun Spice Mix

Salt

6 slices bacon

Olive oil spray

Instructions:

1. Cut the pork in half so that it will fit in the air fryer basket.
   Place both pieces of meat in a resealable plastic bag. Add the oil, Cajun seasoning, and salt to taste, if using. Seal the bag and massage to coat all of the meat with the oil and seasonings. Marinate in the refrigerator for at least 1 hour or up to 24 hours.
2. Remove the pork from the bag and wrap 3 bacon slices around each piece. Spray the Innsky air fryer basket with olive oil spray. Place the meat in the air fryer. Set the Innsky air fryer to 350°F for 15 minutes. Increase the temperature to 400°F for 5 minutes. Use a meat thermometer to ensure the meat has reached an internal temperature of 145°F.
   Let the meat rest for 10 minutes. Slice into 6 medallions and serve.

Panko-Breaded Pork Chops

PREP: 5 MINUTES • COOK TIME: 12 MINUTES • TOTAL: 17 MINUTES • SERVES: 6

Ingredients

5 (3½- to 5-ounce) pork chops (bone-in or boneless)
Seasoning salt
Pepper

¼ cup all-purpose flour
2 tablespoons panko bread crumbs
Cooking oil

Instructions:

1. Season the pork chops with the seasoning salt and pepper to taste.
   Sprinkle the flour on both sides of the pork chops, then coat both sides with panko bread crumbs.
   Place the pork chops in the air fryer. Stacking them is okay.
2. Spray the pork chops with cooking oil. Cook for 6 minutes.
   Open the Air fryer and flip the pork chops. Cook for an additional 6 minutes
   Cool before serving.
   Typically, bone-in pork chops are juicier than boneless. If you prefer really juicy pork chops, use bone-in.

Per Serving: Calories: 246; Fat: 13g; Protein:26g; Fiber:0g

Porchetta-Style Pork Chops

PREP: 10 MINUTES • COOK TIME: 15 MINUTES • TOTAL: 25 MINUTES • SERVES: 2

Ingredients

1 tablespoon extra-virgin olive oil
Grated zest of 1 lemon
2 cloves garlic, minced
2 teaspoons chopped fresh rosemary
1 teaspoon finely chopped fresh sage
1 teaspoon fennel seeds, lightly crushed

¼ to ½ teaspoon red pepper flakes
1 teaspoon kosher salt
1 teaspoon black pepper
(8-ounce) center-cut bone-in pork chops, about 1 inch thick

Instructions:

1. In a small bowl, combine the olive oil, zest, garlic, rosemary, sage, fennel seeds, red pepper, salt, and black pepper. Stir, crushing the herbs with the back of a spoon, until a paste forms. Spread the seasoning mix on both sides of the pork chops.
2. Place the chops in the air fryer basket. Set the Innsky air fryer to 375°F for 15 minutes. Use a meat thermometer to ensure the chops have reached an internal temperature of 145°F.

Apricot Glazed Pork Tenderloins

PREP: 5 MINUTES • COOK TIME: 30 MINUTES • TOTAL: 35 MINUTES • SERVES: 3

Ingredients

1 teaspoon salt
1/2 teaspoon pepper
1-lb pork tenderloin
2 tablespoons minced fresh rosemary or 1 tablespoon dried rosemary, crushed
2 tablespoons olive oil, divided

1 garlic cloves, minced
Apricot Glaze Ingredients
1 cup apricot preserves
3 garlic cloves, minced
4 tablespoons lemon juice

Instructions:

1. Mix well pepper, salt, garlic, oil, and rosemary. Brush all over pork. If needed cut pork crosswise in half to fit in air fryer. Lightly grease baking pan of air fryer with cooking spray. Add pork.
2. For 3 minutes per side, brown pork in a preheated 390°F air fryer.Meanwhile, mix well all glaze Ingredients in a small bowl. Baste pork every 5 minutes. Cook for 20 minutes at 330°F. Serve and enjoy.

Sweet & Spicy Country-Style Ribs
PREP: 10 MINUTES • COOK TIME: 25 MINUTES • TOTAL: 35 MINUTES • SERVES: 4

Ingredients

2 tablespoons brown sugar
2 tablespoons smoked paprika
1 teaspoon garlic powder
1 teaspoon onion powder
1 teaspoon dry mustard
1 teaspoon ground cumin

1 teaspoon kosher salt
1 teaspoon black pepper
¼ to ½ teaspoon cayenne pepper
1½ pounds boneless country-style pork ribs
1 cup barbecue sauce

Instructions:

1. In a small bowl, stir together the brown sugar, paprika, garlic powder, onion powder, dry mustard, cumin, salt, black pepper, and cayenne. Mix until well combined.
   Pat the ribs dry with a paper towel. Generously sprinkle the rub evenly over both sides of the ribs and rub in with your fingers.
2. Place the ribs in the air fryer basket. Set the Innsky air fryer to 350°F for 15 minutes. Turn the ribs and brush with ½ cup of the barbecue sauce. Cook for an additional 10 minutes. Use a meat thermometer to ensure the pork has reached an internal temperature of 145°F. Serve with remaining barbecue sauce.

Pork Tenders With Bell Peppers
PREP: 5 MINUTES • COOK TIME: 15 MINUTES • TOTAL: 20 MINUTES • SERVES: 4

Ingredients

11 Ozs Pork Tenderloin
1 Bell Pepper, in thin strips
1 Red Onion, sliced
2 Tsps Provencal Herbs

Black Pepper to taste
1 Tbsp Olive Oil
1/2 Tbsp Mustard

Instructions:

1. Preheat the Innsky air fryer to 390 degrees.
   In the oven dish, mix the bell pepper strips with the onion, herbs, and some salt and pepper to taste.
   Add half a tablespoon of olive oil to the mixture
   Cut the pork tenderloin into four pieces and rub with salt, pepper and mustard.
   Thinly coat the pieces with remaining olive oil and place them upright in the oven dish on top of the pepper mixture
2. Place the bowl into the Air fryer. Set the timer to 15 minutes and roast the meat and the vegetables
   Turn the meat and mix the peppers halfway through
   Serve with a fresh salad

Wonton Meatballs

PREP: 15 MINUTES • COOK TIME: 10 MINUTES • TOTAL: 25 MINUTES • SERVES: 4

Ingredients

1 pound ground pork
2 large eggs
¼ cup chopped green onions (white and green parts)
¼ cup chopped fresh cilantro or parsley
1 tablespoon minced fresh ginger

3 cloves garlic, minced
2 teaspoons soy sauce
1 teaspoon oyster sauce
½ teaspoon kosher salt
1 teaspoon black pepper

Instructions:

1. In the bowl of a stand mixer fitted with the paddle attachment, combine the pork, eggs, green onions, cilantro, ginger, garlic, soy sauce, oyster sauce, salt, and pepper. Mix on low speed until all of the ingredients are incorporated, 2 to 3 minutes.
   Form the mixture into 12 meatballs and arrange in a single layer in the air fryer basket.
2. Set the Innsky air fryer to 350°F for 10 minutes. Use a meat thermometer to ensure the meatballs have reached an internal temperature of 145°F.
   Transfer the meatballs to a bowl and serve.

Barbecue Flavored Pork Ribs

PREP: 5 MINUTES • COOK TIME: 15 MINUTES • TOTAL: 25 MINUTES • SERVES: 6

Ingredients

¼ cup honey, divided
¾ cup BBQ sauce
2 tablespoons tomato ketchup
1 tablespoon Worcestershire sauce

1 tablespoon soy sauce
½ teaspoon garlic powder
Freshly ground white pepper, to taste
1¾ pound pork ribs

Instructions:

1. In a large bowl, mix together 3 tablespoons of honey and remaining ingredients except pork ribs. Refrigerate to marinate for about 20 minutes. Preheat the Innsky air fryer to 355 degrees F. Place the ribs in an Air fryer basket.
2. Cook for about 13 minutes. Remove the ribs from the Air fryer and coat with remaining honey. Serve hot.

Easy Air Fryer Marinated Pork Tenderloin

PREP:10 MINUTES PLUS 1HOUR TO MARINATE•COOK TIME:30 MINUTES•TOTAL:1 HOUR 40 MINUTES•SERVES:4-6

Ingredients

¼ cup olive oil
¼ cup soy sauce
¼ cup freshly squeezed lemon juice
1 garlic clove, minced

1 tablespoon Dijon mustard
1 teaspoon salt
½ teaspoon freshly ground black pepper
2 pounds pork tenderloin

Instructions:

1. In a large mixing bowl, make the marinade. Mix together the olive oil, soy sauce, lemon juice, minced garlic, Dijon mustard, salt, and pepper. Reserve ¼ cup of the marinade.
   Place the tenderloin in a large bowl and pour the remaining marinade over the meat. Cover and marinate in the refrigerator for about 1 hour. Place the marinated pork tenderloin into the air fryer basket.
2. Set the temperature of your Innsky AF to 400°F. Set the timer and roast for 10 minutes. Using tongs, flip the pork and baste it with half of the reserved marinade. Reset the timer and roast for 10 minutes more.
   Using tongs, flip the pork, then baste with the remaining marinade.
   Reset the timer and roast for another 10 minutes, for a total cooking time of 30 minutes.

Balsamic Glazed Pork Chops

PREP: 5 MINUTES • COOK TIME: 50 MINUTES • TOTAL: 55 MINUTES • SERVES: 4

Ingredients

¾ cup balsamic vinegar
1 ½ tablespoons sugar
1 tablespoon butter

3 tablespoons olive oil
3 tablespoons salt
3 pork rib chops

Instructions:

1   Place all ingredients in bowl and allow the meat to marinate in the fridge for at least 2 hours. Preheat the Innsky air fryer to 390°F. Place the grill pan accessory in the air fryer.

2   Grill the pork chops for 20 minutes making sure to flip the meat every 10 minutes for even grilling. Meanwhile, pour the balsamic vinegar on a saucepan and allow to simmer for at least 10 minutes until the sauce thickens.Brush the meat with the glaze before serving.

Per Serving: Calories: 274; Fat: 18g; Protein:17g

Perfect Air Fried Pork Chops

PREP: 5 MINUTES • COOK TIME: 17 MINUTES • TOTAL: 22 MINUTES • SERVES: 4

Ingredients

3 cups bread crumbs
½ cup grated Parmesan cheese
2 tablespoons vegetable oil
2 teaspoons salt

2 teaspoons sweet paprika
½ teaspoon onion powder
¼ teaspoon garlic powder
6 (½-inch-thick) bone-in pork chops

Instructions:

1   Spray the Innsky air fryer basket with olive oil. In a large resealable bag, combine the bread crumbs, Parmesan cheese, oil, salt, paprika, onion powder, and garlic powder. Seal the bag and shake it a few times in order for the spices to blend together. Place the pork chops, one by one, in the bag and shake to coat.

2   Place the pork chops in the greased Innsky air fryer basket in a single layer. Be careful not to overcrowd the basket. Spray the chops generously with olive oil to avoid powdery, uncooked breading.
Set the temperature of your Innsky AF to 360°F. Set the timer and roast for 10 minutes.
Using tongs, flip the chops. Spray them generously with olive oil.
Reset the timer and roast for 7 minutes more.
Check that the pork has reached an internal temperature of 145°F. Add cooking time if needed.

Per Serving: Calories: 513; Fat: 23g; Saturated fat: 8g; Carbohydrate: 22g; Fiber: 2g; Sugar: 3g; Protein: 50g; Iron: 3mg; Sodium: 1521mg

Rustic Pork Ribs

PREP: 5 MINUTES • COOK TIME: 15 MINUTES • TOTAL: 25 MINUTES • SERVES: 4

Ingredients

1 rack of pork ribs
3 tablespoons dry red wine
1 tablespoon soy sauce
1/2 teaspoon dried thyme
1/2 teaspoon onion powder

1/2 teaspoon garlic powder
1/2 teaspoon ground black pepper
1 teaspoon smoke salt
1 tablespoon cornstarch
1/2 teaspoon olive oil

Instructions:

1.   Begin by preheating your Air fryer to 390 degrees F. Place all ingredients in a mixing bowl and let them marinate at least 1 hour.

2.   Cook the marinated ribs approximately 25 minutes at 390 degrees F. Serve hot.

Air Fryer Baby Back Ribs
PREP: 5 MINUTES • COOK TIME: 25 MINUTES • TOTAL: 30 MINUTES • SERVES: 4

Ingredients
1 rack baby back ribs
1 tablespoon garlic powder
1 teaspoon freshly ground black pepper

2 tablespoons salt
1 cup barbecue sauce (any type)

Instructions:
1. Dry the ribs with a paper towel.
   Season the ribs with the garlic powder, pepper, and salt.
   Place the seasoned ribs into the air fryer.
2. Set the temperature of your Innsky AF to 400°F. Set the timer and grill for 10 minutes.
   Using tongs, flip the ribs.
   Reset the timer and grill for another 10 minutes.
   Once the ribs are cooked, use a pastry brush to brush on the barbecue sauce, then set the timer and grill for a final 3 to 5 minutes.

Per Serving: Calories: 422; Fat: 27g; Saturated fat: 10g; Carbohydrate: 25g; Fiber: 1g; Sugar: 17g; Protein: 18g; Iron: 1mg; Sodium: 4273mg

Keto Parmesan Crusted Pork Chops
PREP: 10 MINUTES • COOK TIME: 15 MINUTES • TOTAL: 25 MINUTES • SERVES: 8

Ingredients
3 tbsp. grated parmesan cheese
1 C. pork rind crumbs
2 beaten eggs
¼ tsp. chili powder
½ tsp. onion powder

1 tsp. smoked paprika
¼ tsp. pepper
½ tsp. salt
4-6 thick boneless pork chops

Instructions:
1. Ensure your air fryer is preheated to 400 degrees.
   With pepper and salt, season both sides of pork chops.
   In a food processor, pulse pork rinds into crumbs. Mix crumbs with other seasonings.
   Beat eggs and add to another bowl.
   Dip pork chops into eggs then into pork rind crumb mixture.
2. Spray down air fryer with olive oil and add pork chops to the basket. Set temperature to 400°F, and set time to 15 minutes.

Per Serving: Calories: 422; Fat: 19g; Protein:38g; Sugar:2g

Pork Milanese
PREP: 10 MINUTES • COOK TIME: 12 MINUTES • TOTAL: 22 MINUTES • SERVES: 4

Ingredients
4 (1-inch) boneless pork chops
Fine sea salt and ground black pepper
2 large eggs

¾ cup powdered Parmesan cheese about 2¼ ounces
Chopped fresh parsley, for garnish
Lemon slices, for serving

Instructions:

1. Spray the Innsky air fryer basket with avocado oil. Preheat the Innsky air fryer to 400°F. Place the pork chops between 2 sheets of plastic wrap and pound them with the flat side of a meat tenderizer until they're ¼ inch thick. Lightly season both sides of the chops with salt and pepper. Lightly beat the eggs in a shallow bowl. Divide the Parmesan cheese evenly between 2 bowls and set the bowls in this order: Parmesan, eggs, Parmesan. Dredge a chop in the first bowl of Parmesan, then dip it in the eggs, and then dredge it again in the second bowl of Parmesan, making sure both sides and all edges are well coated. Repeat with the remaining chops.

2. Place the chops in the Innsky air fryer basket and cook for 12 minutes, or until the internal temperature reaches 145°F, flipping halfway through.

   Garnish with fresh parsley and serve immediately with lemon slices. Store leftovers in an airtight container in the refrigerator for up to 3 days. Reheat in a preheated 390°F air fryer for 5 minutes, or until warmed through.

Per serving: Calories **351;** Fat **18g;** Protein **42g;** Total carbs **3g;** Fiber **1g**

Crispy Fried Pork Chops the Southern Way
PREP: 10 MINUTES • COOK TIME: 25 MINUTES • TOTAL: 35 MINUTES • SERVES: 4

Ingredients
½ cup all-purpose flour
½ cup low fat buttermilk
½ teaspoon black pepper

½ teaspoon Tabasco sauce
teaspoon paprika
3 bone-in pork chops

Instructions:

1. Place the buttermilk and hot sauce in a Ziploc bag and add the pork chops. Allow to marinate for at least an hour in the fridge.

   In a bowl, combine the flour, paprika, and black pepper.

   Remove pork from the Ziploc bag and dredge in the flour mixture.

   Preheat the Innsky air fryer to 390°F.

   Spray the pork chops with cooking oil.

2. Place in the Innsky air fryer basket and cook for 25 minutes.

Per Serving: Calories: 427; Fat: 21.2g; Protein:46.4g; Sugar:2g

Italian Sausages with Peppers and Onions
PREP: 5 MINUTES • COOK TIME: 28 MINUTES • TOTAL: 33 MINUTES • SERVES: 3

Ingredients

1 medium onion, thinly sliced
1 yellow or orange bell pepper, thinly sliced
1 red bell pepper, thinly sliced
¼ cup avocado oil or melted coconut oil

1 teaspoon fine sea salt
6 Italian sausages
Dijon mustard, for serving (optional)

Instructions:

1   Preheat the Innsky air fryer to 400°F. Place the onion and peppers in a large bowl. Drizzle with the oil and toss well to coat the veggies. Season with the salt.
    Place the onion and peppers in a 6-inch pie pan and cook in the air fryer for 8 minutes, stirring halfway through. Remove from the air fryer and set aside.

2   Spray the Innsky air fryer basket with avocado oil. Place the sausages in the Innsky air fryer basket and cook for 20 minutes, or until crispy and golden brown. During the last minute or two of cooking, add the onion and peppers to the basket with the sausages to warm them through.
    Place the onion and peppers on a serving platter and arrange the sausages on top. Serve Dijon mustard on the side, if desired.
    Store leftovers in an airtight container in the fridge for up to 7 days or in the freezer for up to a month. Reheat in a preheated 390°F air fryer for 3 minutes, or until heated through.

Per serving: Calories 576; Fat 49g; Protein 25g; Total carbs 8g; Fiber 2g

Fried Pork Quesadilla
PREP: 10 MINUTES • COOK TIME: 12 MINUTES • TOTAL: 22 MINUTES • SERVES: 2

Ingredients

Two 6-inch corn or flour tortilla shells
1 medium-sized pork shoulder, approximately 4 ounces, sliced
½ medium-sized white onion, sliced
½ medium-sized red pepper, sliced

½ medium sized green pepper, sliced
½ medium sized yellow pepper, sliced
¼ cup of shredded pepper-jack cheese
¼ cup of shredded mozzarella cheese

Instructions:

1   Preheat the Innsky air fryer to 350 degrees.
    In the oven on high heat for 20 minutes, grill the pork, onion, and peppers in foil in the same pan, allowing the moisture from the vegetables and the juice from the pork mingle together. Remove pork and vegetables in foil from the oven. While they're cooling, sprinkle half the shredded cheese over one of the tortillas, then cover with the pieces of pork, onions, and peppers, and then layer on the rest of the shredded cheese. Top with the second tortilla. Place directly on hot surface of the Air fryer basket.

2   Set the Innsky air fryer timer for 6 minutes. After 6 minutes, when the Air fryer shuts off, flip the tortillas onto the other side with a spatula; the cheese should be melted enough that it won't fall apart, but be careful anyway not to spill any toppings!
    ReSet the Innsky air fryer to 350 degrees for another 6 minutes.
    After 6 minutes, when the air fryer shuts off, the tortillas should be browned and crisp, and the pork, onion, peppers and cheese will be crispy and hot and delicious. Remove with tongs and let sit on a serving plate to cool for a few minutes before slicing.

Cilantro-Mint Pork BBQ Thai Style
PREP: 5 MINUTES • COOK TIME: 15 MINUTES • TOTAL: 20 MINUTES • SERVES: 3

Ingredients
1 minced hot chile
1 minced shallot
1-pound ground pork
2 tablespoons fish sauce

2 tablespoons lime juice
3 tablespoons basil
3 tablespoons chopped mint
3 tablespoons cilantro

Instructions:
1   In a shallow dish, mix well all Ingredients with hands. Form into 1-inch ovals. Thread ovals in skewers. Place on skewer rack in air fryer.
2   For 15 minutes, cook on 360°F. Halfway through cooking time, turnover skewers. If needed, cook in batches. Serve and enjoy.

Per Serving: Calories: 455; Fat: 31.5g; Protein:40.4g

Pork Wonton Wonderful
PREP: 10 MINUTES • COOK TIME: 25 MINUTES • TOTAL: 35 MINUTES • SERVES: 3

Ingredients
8 wanton wrappers
4 ounces of raw minced pork
1 medium-sized green apple
1 cup of water, for wetting the wanton wrappers

1 tablespoon of vegetable oil
½ tablespoon of oyster sauce
1 tablespoon of soy sauce
Large pinch of ground white pepper

Instructions:
1   Cover the basket of the Air fryer with a lining of tin foil, leaving the edges uncovered to allow air to circulate through the basket. Preheat the Innsky air fryer to 350 degrees. In a small mixing bowl, combine the oyster sauce, soy sauce, and white pepper, then add in the minced pork and stir thoroughly. Cover and set in the fridge to marinate for at least 15 minutes. Core the apple, and slice into small cubes – smaller than bite-sized chunks. Add the apples to the marinating meat mixture, and combine thoroughly. Spread the wonton wrappers, and fill each with a large spoonful of the filling. Wrap the wontons into triangles, so that the wrappers fully cover the filling, and seal with a drop of the water. Coat each filled and wrapped wonton thoroughly with the vegetable oil, to help ensure a nice crispy fry. Place the wontons on the foil-lined air-fryer basket.
2   Set the Innsky air fryer timer to 25 minutes. Halfway through cooking time, shake the handle of the Innsky air fryer basket vigorously to jostle the wontons and ensure even frying. After 25 minutes, when the Air fryer shuts off, the wontons will be crispy golden-brown on the outside and juicy and delicious on the inside. Serve directly from the Innsky air fryer basket and enjoy while hot.

Crispy Roast Garlic-Salt Pork
PREP: 5 MINUTES • COOK TIME: 45 MINUTES • TOTAL: 50 MINUTES • SERVES: 4

Ingredients
1 teaspoon Chinese five spice powder
1 teaspoon white pepper

2 pounds pork belly
2 teaspoons garlic salt

Instructions:

1   Preheat the Innsky air fryer to 390°F.
Mix all the spices in a bowl to create the dry rub.
Score the skin of the pork belly with a knife and season the entire pork with the spice rub.
2   Place in the Innsky air fryer basket and cook for 40 to 45 minutes until the skin is crispy.
Chop before serving.

Scotch Eggs

PREP: 10 MINUTES • COOK TIME: 15 MINUTES • TOTAL: 25 MINUTES • SERVES: 8

Ingredients

2 pounds ground pork or ground beef
2 teaspoons fine sea salt
½ teaspoon ground black pepper, plus more for garnish

8 large hard-boiled eggs, peeled
2 cups pork dust
Dijon mustard, for serving (optional)

Instructions:

Spray the Innsky air fryer basket with avocado oil. Preheat the Innsky air fryer to 400°F.Place the ground pork in a large bowl, add the salt and pepper, and use your hands to mix until seasoned throughout. Flatten about ¼ pound of ground pork in the palm of your hand and place a peeled egg in the center. Fold the pork completely around the egg. Repeat with the remaining eggs. Place the pork dust in a medium-sized bowl. One at a time, roll the ground pork–covered eggs in the pork dust and use your hands to press it into the eggs to form a nice crust.

Place the eggs in the Innsky air fryer basket and spray them with avocado oil.

Cook the eggs for 15 minutes, or until the internal temperature of the pork reaches 145°F and the outside is golden brown. Garnish with ground black pepper and serve with Dijon mustard, if desired.

Store leftovers in an airtight container in the fridge for up to 7 days or in the freezer for up to a month. Reheat in a preheated 400°F air fryer for 3 minutes, or until heated through.

Per serving: Calories 447; Fat 34g; Protein 43g; Total carbs 0.5g; Fiber 0g

Italian Parmesan Breaded Pork Chops

PREP: 5 MINUTES • COOK TIME: 25 MINUTES • TOTAL: 30 MINUTES • SERVES: 5

Ingredients

5 (3½- to 5-ounce) pork chops (bone-in or boneless)
1 teaspoon Italian seasoning
Seasoning salt
Pepper

¼ cup all-purpose flour
2 tablespoons Italian bread crumbs
3 tablespoons finely grated Parmesan cheese
Cooking oil

Instructions:

1   Season the pork chops with the Italian seasoning and seasoning salt and pepper to taste. Sprinkle the flour on both sides of the pork chops, then coat both sides with the bread crumbs and Parmesan cheese.

2   Place the pork chops in the Air fryer. Stacking them is okay. Spray the pork chops with cooking oil. Cook for 6 minutes. Open the Air fryer and flip the pork chops. Cook for an additional 6 minutes. Cool before serving. Instead of seasoning salt, you can use either chicken or pork rub for additional flavor. You can find these rubs in the spice aisle of the grocery store.

Per Serving: Calories: 334; Fat: 7g; Protein:34g; Fiber:0g

Pork Tenderloin with Avocado Lime Sauce

PREP:10 MINUTES PLUS 2 HOURS TO MARINATE•COOK TIME:15 MINUTES•TOTAL: 2 HOURS 25MINUTES• SERVES:4

Ingredients

MARINADE:
½ cup lime juice
Grated zest of 1 lime
2 teaspoons stevia glycerite, or ¼ teaspoon liquid stevia
3 cloves garlic, minced
1½ teaspoons fine sea salt
1 teaspoon chili powder, or more for more heat
1 teaspoon smoked paprika
1 pound pork tenderloin
AVOCADO LIME SAUCE:

1 medium-sized ripe avocado, roughly chopped
½ cup full-fat sour cream
Grated zest of 1 lime
Juice of 1 lime
2 cloves garlic, roughly chopped
½ teaspoon fine sea salt
¼ teaspoon ground black pepper
Chopped fresh cilantro leaves, for garnish
Lime slices, for serving
Pico de gallo, for serving

Instructions:

1 In a medium-sized casserole dish, stir together all the marinade ingredients until well combined. Add the tenderloin and coat it well in the marinade. Cover and place in the fridge to marinate for 2 hours or overnight. Spray the Innsky air fryer basket with avocado oil. Preheat the Innsky air fryer to 400°F.

2 Remove the pork from the marinade and place it in the air fryer basket. Cook for 13 to 15 minutes, until the internal temperature of the pork is 145°F, flipping after 7 minutes. Remove the pork from the air fryer and place it on a cutting board. Allow it to rest for 8 to 10 minutes, then cut it into ½-inch-thick slices.

While the pork cooks, make the avocado lime sauce: Place all the sauce ingredients in a food processor and puree until smooth. Taste and adjust the seasoning to your liking.

Place the pork slices on a serving platter and spoon the avocado lime sauce on top. Garnish with cilantro leaves and serve with lime slices and pico de gallo.

Store leftovers in an airtight container in the fridge for up to 4 days. Reheat in a preheated 400°F air fryer for 5 minutes, or until heated through.

Per serving: Calories 326; Fat 19g; Protein 26g; Total carbs 15g; Fiber 6g

Tuscan Pork Chops

PREP: 10 MINUTES • COOK TIME: 10 MINUTES • TOTAL: 20 MINUTES • SERVES: 4

Ingredients

1/4 cup all-purpose flour
1 teaspoon salt
3/4 teaspoons seasoned pepper
4 (1-inch-thick) boneless pork chops
1 tablespoon olive oil

3 to 4 garlic cloves
1/3 cup balsamic vinegar
1/3 cup chicken broth
3 plum tomatoes, seeded and diced
2 tablespoons capers

Instructions:

Combine flour, salt, and pepper

Press pork chops into flour mixture on both sides until evenly covered.

Cook in your Air fryer at 360 degrees for 14 minutes, flipping half way through.

While the pork chops cook, warm olive oil in a medium skillet. Add garlic and sauté for 1 minute; then mix in vinegar and chicken broth. Add capers and tomatoes and turn to high heat. Bring the sauce to a boil, stirring regularly, then add pork chops, cooking for one minute. Remove from heat and cover for about 5 minutes to allow the pork to absorb some of the sauce; serve hot.

Per Serving: Calories: 349; Fat: 23g; Protein:20g; Fiber:1.5g

BBQ Riblets

PREP: 10 MINUTES • COOK TIME: 25 MINUTES • TOTAL: 35 MINUTES • SERVES: 4

Ingredients

1 rack pork riblets, cut into individual riblets
1 teaspoon fine sea salt
1 teaspoon ground black pepper
SAUCE:
¼ cup apple cider vinegar
¼ cup beef broth

¼ cup Swerve confectioners'-style sweetener or powdered sweetener
¼ cup tomato sauce
1 teaspoon liquid smoke
1 teaspoon onion powder
2 cloves garlic, minced

Instructions:

1 Spray the Innsky air fryer basket with avocado oil. Preheat the Innsky air fryer to 350°F. Season the riblets well on all sides with the salt and pepper.

2 Place the riblets in the Innsky air fryer basket and cook for 10 minutes, flipping halfway through.

   While the riblets cook, mix all the sauce ingredients together in a 6-inch pie pan.

   Remove the riblets from the air fryer and place them in the pie pan with the sauce. Stir to coat the riblets in the sauce. Transfer the pan to the air fryer and cook for 10 to 15 minutes, until the pork is cooked through and the internal temperature reaches 145°F.

   Store leftovers in an airtight container in the refrigerator for up to 4 days. Reheat in a preheated 350°F air fryer for 5 minutes, or until heated through.

Per serving: Calories 319; Fat 26g; Protein 19g; Total carbs 3g; Fiber 0.3

Five-Spice Pork Belly

PREP: 10 MINUTES • COOK TIME: 17 MINUTES • TOTAL: 27 MINUTES • SERVES: 4

Ingredients

1 pound unsalted pork belly
2 teaspoons Chinese five-spice powder
SAUCE:
1 tablespoon coconut oil
1 (1-inch) piece fresh ginger, peeled and grated
2 cloves garlic, minced
½ cup beef or chicken broth

¼ to ½ cup Swerve confectioners'-style sweetener
3 tablespoons wheat-free tamari, or ½ cup coconut aminos
1 green onion, sliced, plus more for garnish
1 drop orange oil, or ½ teaspoon orange extract (optional)

Instructions:

1   Spray the Innsky air fryer basket with avocado oil. Preheat the Innsky air fryer to 400°F. Cut the pork belly into ½-inch-thick slices and season well on all sides with the five-spice powder.

2   Place the slices in a single layer in the Innsky air fryer basket (if you're using a smaller air fryer, work in batches if necessary) and cook for 8 minutes, or until cooked to your liking, flipping halfway through. While the pork belly cooks, make the sauce: Heat the coconut oil in a small saucepan over medium heat. Add the ginger and garlic and sauté for 1 minute, or until fragrant. Add the broth, sweetener, and tamari and simmer for 10 to 15 minutes, until thickened. Add the green onion and cook for another minute, until the green onion is softened. Add the orange oil (if using). Taste and adjust the seasoning to your liking.

Transfer the pork belly to a large bowl. Pour the sauce over the pork belly and coat well. Place the pork belly slices on a serving platter and garnish with sliced green onions.

Best served fresh. Store leftovers in an airtight container in the fridge for up to 4 days. Reheat in a preheated 400°F air fryer for 3 minutes, or until heated through.

Per serving: Calories 365; Fat 32g; Protein 19g; Total carbs 2g; Fiber 0.3g

Caramelized Pork Shoulder

PREP: 10 MINUTES • COOK TIME: 20 MINUTES • TOTAL: 30 MINUTES • SERVES: 8

Ingredients

1/3 cup soy sauce
2 tablespoons sugar

1 tablespoon honey
3 pound pork shoulder, cut into 1½-inch thick slices

Instructions:

1   In a bowl, mix together all ingredients except pork.
Add pork and coat with marinade generously. Cover and refrigerate o marinate for about 2-8 hours. Preheat the Innsky air fryer to 335 degrees F.

2   Place the pork in an Air fryer basket. Cook for about 10 minutes.
Now, Set the Innsky air fryer to 390 degrees F. Cook for about 10 minutes

Ginger, Garlic And Pork Dumplings
PREP: 10 MINUTES • COOK TIME: 15 MINUTES • TOTAL: 25 MINUTES • SERVES: 8

Ingredients

¼ teaspoon crushed red pepper
½ teaspoon sugar
1 tablespoon chopped fresh ginger
1 tablespoon chopped garlic
1 teaspoon canola oil
1 teaspoon toasted sesame oil

18 dumpling wrappers
2 tablespoons rice vinegar
2 teaspoons soy sauce
4 cups bok choy, chopped
4 ounces ground pork

Instructions:

1. Heat oil in a skillet and sauté the ginger and garlic until fragrant. Stir in the ground pork and cook for 5 minutes. Stir in the bok choy and crushed red pepper. Season with salt and pepper to taste. Allow to cool. Place the meat mixture in the middle of the dumpling wrappers. Fold the wrappers to seal the meat mixture in.
    Place the bok choy in the grill pan.
2. Cook the dumplings in the air fryer at 330°F for 15 minutes.
    Meanwhile, prepare the dipping sauce by combining the remaining Ingredients in a bowl.

Per Serving: Calories: 137; Fat: 5g; Protein:7g

Peanut Satay Pork
PREP: 5 MINUTES • COOK TIME: 12 MINUTES • TOTAL: 17 MINUTES • SERVES: 5

Ingredients

11 Ozs Pork Fillet, sliced into bite sized strips
4 Cloves Garlic, crushed
1 Tsp Ginger Powder
2 Tsps Chili Paste
2 Tbsps Sweet Soy Sauce

2 Tbsps Vegetable Oil
1 Shallot, finely chopped
1 Tsp Ground Coriander
3/4 Cup Coconut Milk
1/3 Cup Peanuts, ground

Instructions:

1   Mix half of the garlic in a dish with the ginger, a tablespoon of sweet soy sauce, and a tablespoon of the oil. Combine the meat into the mixture and leave to marinate for 15 minutes. Preheat the Innsky air fryer to 390 degrees
2   Place the marinated meat into the Air fryer. Set the timer to 12 minutes and roast the meat until brown and done. Turn once while roasting
    In the meantime, make the peanut sauce by heating the remaining tablespoon of oil in a saucepan and gently sauté the shallot with the garlic. Add the coriander and fry until fragrant
    Mix the coconut milk and the peanuts with the chili paste and remaining soy sauce with the shallot mixture and gently boil for 5 minutes, while stirring
    Drizzle over the cooked meat and serve with rice.

Dry Rub Baby Back Ribs

PREP: 5 MINUTES • COOK TIME: 35 MINUTES • TOTAL: 40 MINUTES • SERVES: 2

Ingredients

2 teaspoons fine sea salt
1 teaspoon ground black pepper
2 teaspoons smoked paprika
1 teaspoon garlic powder

1 teaspoon onion powder
½ teaspoon chili powder (optional)
1 rack baby back ribs, cut in half crosswise

Instructions:

1 Spray the Innsky air fryer basket with avocado oil. Preheat the Innsky air fryer to 350°F. In a small bowl, combine the salt, pepper, and seasonings. Season the ribs on all sides with the seasoning mixture.
2 Place the ribs in the Innsky air fryer basket and cook for 15 minutes, then flip the ribs over and cook for another 15 to 20 minutes, until the ribs are cooked through and the internal temperature reaches 145°F.

Per serving: Calories 515, Fat 40g; Protein 37g; Total carbs 3g; Fiber 1g

Crispy Breaded Pork Chops

PREP: 10 MINUTES • COOK TIME: 15 MINUTES • TOTAL: 25 MINUTES • SERVES: 8

Ingredients

1/8 tsp. pepper
¼ tsp. chili powder
½ tsp. onion powder
½ tsp. garlic powder
1 ¼ tsp. sweet paprika

2 tbsp. grated parmesan cheese
1/3 C. crushed cornflake crumbs
½ C. panko breadcrumbs
1 beaten egg
6 center-cut boneless pork chops

Instructions:

1 Ensure that your air fryer is preheated to 400 degrees. Spray the basket with olive oil.
   With ½ teaspoon salt and pepper, season both sides of pork chops.
   Combine ¾ teaspoon salt with pepper, chili powder, onion powder, garlic powder, paprika, cornflake crumbs, panko breadcrumbs and parmesan cheese.
   Beat egg in another bowl.
   Dip pork chops into the egg and then crumb mixture.
   Add pork chops to air fryer and spritz with olive oil.
2 Set temperature to 400°F, and set time to 12 minutes. Cook 12 minutes, making sure to flip over halfway through cooking process.
   Only add 3 chops in at a time and repeat the process with remaining pork chops.

Per Serving: Calories: 378; Fat: 13g; Protein:33g; Sugar:1g

Bacon-Wrapped Stuffed Pork Chops
PREP: 10 MINUTES • COOK TIME: 20 MINUTES • TOTAL: 30 MINUTES • SERVES: 4

Ingredients
4 (1-inch-thick) boneless pork chops

2 (5.2-ounce) packages Boursin cheese

8 slices thin-cut bacon

Instructions:

1   Spray the Innsky air fryer basket with avocado oil. Preheat the Innsky air fryer to 400°F. Place one of the chops on a cutting board. With a sharp knife held parallel to the cutting board, make a 1-inch-wide incision on the top edge of the chop. Carefully cut into the chop to form a large pocket, leaving a ½-inch border along the sides and bottom. Repeat with the other 3 chops. Snip the corner of a large resealable plastic bag to form a ¾-inch hole. Place the Boursin cheese in the bag and pipe the cheese into the pockets in the chops, dividing the cheese evenly among them.

Wrap 2 slices of bacon around each chop and secure the ends with toothpicks.

2   Place the bacon-wrapped chops in the Innsky air fryer basket and cook for 10 minutes, then flip the chops and cook for another 8 to 10 minutes, until the bacon is crisp, the chops are cooked through, and the internal temperature reaches 145°F.

Store leftovers in an airtight container in the refrigerator for up to 3 days. Reheat in a preheated 400°F air fryer for 5 minutes, or until warmed through.

Per serving: Calories 578; Fat 45g; Protein 37g; Total carbs 16g; Fiber 1g

Curry Pork Roast in Coconut Sauce
PREP: 10 MINUTES • COOK TIME: 60 MINUTES • TOTAL: 70 MINUTES • SERVES: 6

Ingredients
½ teaspoon curry powder

½ teaspoon ground turmeric powder

1 can unsweetened coconut milk

1 tablespoons sugar

2 tablespoons fish sauce

2 tablespoons soy sauce

3 pounds pork shoulder

Salt and pepper to taste

Instructions:

1   Place all Ingredients in bowl and allow the meat to marinate in the fridge for at least 2 hours.

Preheat the Innsky air fryer to 390°F.

Place the grill pan accessory in the air fryer.

2   Grill the meat for 20 minutes making sure to flip the pork every 10 minutes for even grilling and cook in batches.

Meanwhile, pour the marinade in a saucepan and allow to simmer for 10 minutes until the sauce thickens.

Baste the pork with the sauce before serving.

Per Serving: Calories: 688; Fat: 52g; Protein:17g

Chinese Salt and Pepper Pork Chop Stir-fry
PREP: 10 MINUTES • COOK TIME: 15 MINUTES • TOTAL: 25 MINUTES • SERVES: 4

Ingredients

Pork Chops:
Olive oil
¾ C. almond flour
¼ tsp. pepper
½ tsp. salt
1 egg white
Pork Chops

Stir-fry:
¼ tsp. pepper
1 tsp. sea salt
2 tbsp. olive oil
2 sliced scallions
2 sliced jalapeno peppers

Instructions:

1   Coat the Innsky air fryer basket with olive oil. Whisk pepper, salt, and egg white together till foamy. Cut pork chops into pieces, leaving just a bit on bones. Pat dry. Add pieces of pork to egg white mixture, coating well. Let sit for marinade 20 minutes. Put marinated chops into a large bowl and add almond flour. Dredge and shake off excess and place into air fryer.

2   Set temperature to 360°F, and set time to 12 minutes. Cook 12 minutes at 360 degrees. Turn up the heat to 400 degrees and cook another 6 minutes till pork chops are nice and crisp. To make stir-fry, remove jalapeno seeds and chop up. Chop scallions and mix with jalapeno pieces. Heat a skillet with olive oil. Stir-fry pepper, salt, scallions, and jalapenos 60 seconds. Then add fried pork pieces to skills and toss with scallion mixture. Stir-fry 1-2 minutes till well coated and hot.

Per Serving: Calories: 294; Fat: 17g; Protein:36g; Sugar:4g

Roasted Pork Tenderloin
PREP: 5 MINUTES • COOK TIME: 1 HOUR • TOTAL: 65 MINUTES • SERVES: 4

Ingredients

1 (3-pound) pork tenderloin
2 tablespoons extra-virgin olive oil
2 garlic cloves, minced
1 teaspoon dried basil

1 teaspoon dried oregano
1 teaspoon dried thyme
Salt
Pepper

Instructions:

1   Drizzle the pork tenderloin with the olive oil.
Rub the garlic, basil, oregano, thyme, and salt and pepper to taste all over the tenderloin.

2   Place the tenderloin in the Air fryer. Cook for 45 minutes.
Use a meat thermometer to test for doneness
Open the Air fryer and flip the pork tenderloin. Cook for an additional 15 minutes.
Remove the cooked pork from the air fryer and allow it to rest for 10 minutes before cutting.

Garlic Putter Pork Chops
PREP: 10 MINUTES • COOK TIME: 7 MINUTES • TOTAL: 17 MINUTES • SERVES: 4

Ingredients

2 tsp. parsley
2 tsp. grated garlic cloves
1 tbsp. coconut oil

1 tbsp. coconut butter
4 pork chops

Instructions:

1   Ensure your air fryer is preheated to 350 degrees.
Mix butter, coconut oil, and all seasoning together. Then rub seasoning mixture over all sides of pork chops. Place in foil, seal, and chill for 1 hour. Remove pork chops from foil and place into air fryer.

2   Set temperature to 350°F, and set time to 7 minutes. Cook 7 minutes on one side and 8 minutes on the other. Drizzle with olive oil and serve alongside a green salad.

Per Serving: Calories: 526; Fat: 23g; Protein:41g; Sugar:4g

Fried Pork with Sweet and Sour Glaze
PREP: 5 MINUTES • COOK TIME: 30 MINUTES • TOTAL: 35 MINUTES • SERVES: 4

Ingredients

¼ cup rice wine vinegar
¼ teaspoon Chinese five spice powder
1 cup potato starch
1 green onion, chopped
2 large eggs, beaten

2 pounds pork chops cut into chunks
2 tablespoons cornstarch + 3 tablespoons water
5 tablespoons brown sugar
Salt and pepper to taste

Instructions:

1   Preheat the Innsky air fryer to 390°F. Season pork chops with salt and pepper to taste. Dip the pork chops in egg. Set aside. In a bowl, combine the potato starch and Chinese five spice powder. Dredge the pork chops in the flour mixture.

2   Place in the double layer rack and cook for 30 minutes. Meanwhile, place the vinegar and brown sugar in a saucepan. Season with salt and pepper to taste. Stir in the cornstarch slurry and allow to simmer until thick. Serve the pork chops with the sauce and garnish with green onions.

Per Serving: Calories: 420; Fat: 11.8g; Protein:69.2g

Pork Cutlet Rolls
PREP: 10 MINUTES • COOK TIME: 15 MINUTES • TOTAL: 25 MINUTES • SERVES: 4

Ingredients

4 Pork Cutlets
4 Sundried Tomatoes in oil
2 Tbsps Parsley, finely chopped
1 Green Onion, finely chopped

Black Pepper to taste
2 Tsps Paprika
1/2 Tbsp Olive Oil
String for Rolled Meat

Instructions:

1   Preheat the Innsky air fryer to 390 degrees. Finely chop the tomatoes and mix with the parsley and green onion. Add salt and pepper to taste. Spread out the cutlets and coat them with the tomato mixture. Roll up the cutlets and secure intact with the string. Rub the rolls with salt, pepper, and paprika powder and thinly coat them with olive oil

2   Put the cutlet rolls in the Air fryer tray and cook for 15 minutes. Roast until nicely brown and done. Serve with tomato sauce.

Oregano-Paprika on Breaded Pork
PREP: 10 MINUTES • COOK TIME: 30 MINUTES • TOTAL: 40 MINUTES • SERVES: 4

Ingredients

¼ cup water
¼ teaspoon dry mustard
½ teaspoon black pepper
½ teaspoon cayenne pepper
½ teaspoon garlic powder
½ teaspoon salt

1 cup panko breadcrumbs
1 egg, beaten
2 teaspoons oregano
4 lean pork chops
4 teaspoons paprika

Instructions:

1   Preheat the Innsky air fryer to 390°F.

Pat dry the pork chops. In a mixing bowl, combine the egg and water. Then set aside.

In another bowl, combine the rest of the Ingredients. Dip the pork chops in the egg mixture and dredge in the flour mixture.

2   Place in the Innsky air fryer basket and cook for 25 to 30 minutes until golden.

Per Serving: Calories: 364; Fat: 20.2g; Protein:42.9g

Bacon Wrapped Pork Tenderloin
PREP: 5 MINUTES • COOK TIME: 15 MINUTES • TOTAL: 20 MINUTES • SERVES: 4

Ingredients

Pork:

1-2 tbsp. Dijon mustard

3-4 strips of bacon

1 pork tenderloin

Apple Gravy:

½ - 1 tsp. Dijon mustard

1 tbsp. almond flour

2 tbsp. ghee

1 chopped onion

2-3 Granny Smith apples

1 C. vegetable broth

Instructions:

1   Spread Dijon mustard all over tenderloin and wrap meat with strips of bacon.

2   Place into the Air fryer, set temperature to 360°F, and set time to 15 minutes and cook 10-15 minutes at 360 degrees. Use a meat thermometer to check for doneness. To make sauce, heat ghee in a pan and add shallots. Cook 1-2 minutes. Then add apples, cooking 3-5 minutes until softened. Add flour and ghee to make a roux. Add broth and mustard, stirring well to combine. When sauce starts to bubble, add 1 cup of sautéed apples, cooking till sauce thickens. Once pork tenderloin I cook, allow to sit 5-10 minutes to rest before slicing. Serve topped with apple gravy.

Per Serving: Calories: 552; Fat: 25g; Protein:29g; Sugar:6g

Pork Tenders With Bell Peppers
PREP: 5 MINUTES • COOK TIME: 15 MINUTES • TOTAL: 20 MINUTES • SERVES: 4

Ingredients

11 Ozs Pork Tenderloin

1 Bell Pepper, in thin strips

1 Red Onion, sliced

2 Tsps Provencal Herbs

Black Pepper to taste

1 Tbsp Olive Oil

1/2 Tbsp Mustard

Round Oven Dish

Instructions:

1   Preheat the Innsky air fryer to 390 degrees. In the oven dish, mix the bell pepper strips with the onion, herbs, and some salt and pepper to taste. Add half a tablespoon of olive oil to the mixture. Cut the pork tenderloin into four pieces and rub with salt, pepper and mustard. Thinly coat the pieces with remaining olive oil and place them upright in the oven dish on top of the pepper mixture

2   Place the bowl into the Air fryer. Set the timer to 15 minutes and roast the meat and the vegetables. Turn the meat and mix the peppers halfway through. Serve with a fresh salad.

Dijon Garlic Pork Tenderloin
PREP: 5 MINUTES • COOK TIME: 10 MINUTES • TOTAL: 15 MINUTES • SERVES: 6

Ingredients

1 C. breadcrumbs

Pinch of cayenne pepper

3 crushed garlic cloves

2 tbsp. ground ginger

2 tbsp. Dijon mustard

2 tbsp. raw honey

4 tbsp. water

2 tsp. salt

1pound pork tenderloin, sliced into 1-inch rounds

Instructions:

With pepper and salt, season all sides of tenderloin.

Combine cayenne pepper, garlic, ginger, mustard, honey, and water until smooth.

Dip pork rounds into honey mixture and then into breadcrumbs, ensuring they all get coated well. Place coated pork rounds into your Air fryer.

Set temperature to 400°F, and set time to 10 minutes. Cook 10 minutes at 400 degrees. Flip and then cook an additional 5 minutes until golden in color.

Pork Neck with Salad
PREP: 10 MINUTES • COOK TIME: 12 MINUTES • TOTAL: 22 MINUTES • SERVES: 2

Ingredients

For Pork:
1 tablespoon soy sauce
1 tablespoon fish sauce
½ tablespoon oyster sauce
½ pound pork neck
For Salad:
1 ripe tomato, sliced tickly
8-10 Thai shallots, sliced
1 scallion, chopped

1 bunch fresh basil leaves
1 bunch fresh cilantro leaves
For Dressing:
3 tablespoons fish sauce
2 tablespoons olive oil
1 teaspoon apple cider vinegar
1 tablespoon palm sugar
2 bird eye chili
1 tablespoon garlic, minced

Instructions:

1. For pork in a bowl, mix together all ingredients except pork. Add pork neck and coat with marinade evenly. Refrigerate for about 2-3 hours. Preheat the Innsky air fryer to 340 degrees F.
2. Place the pork neck onto a grill pan. Cook for about 12 minutes. Meanwhile in a large salad bowl, mix together all salad ingredients. In a bowl, add all dressing ingredients and beat till well combined. Remove pork neck from Air fryer and cut into desired slices. Place pork slices over salad.

Cajun Pork Steaks
PREP: 5 MINUTES • COOK TIME: 20 MINUTES • TOTAL: 25 MINUTES • SERVES: 6

Ingredients

4-6 pork steaks
BBQ sauce:
Cajun seasoning
1 tbsp. vinegar

1 tsp. low-sodium soy sauce
½ C. brown sugar
½ C. vegan ketchup

Instructions:

1. Ensure your air fryer is preheated to 290 degrees.
   Sprinkle pork steaks with Cajun seasoning.
   Combine remaining ingredients and brush onto steaks. Add coated steaks to air fryer.
2. Set temperature to 290°F, and set time to 20 minutes. Cook 15-20 minutes till just browned.

Per Serving: Calories: 209; Fat: 11g; Protein:28g; Sugar:2g

Wonton Taco Cups
PREP: 5 MINUTES • COOK TIME: 10 MINUTES • TOTAL: 15 MINUTES • SERVES: 8

Ingredients

1/2pound ground pork, browned
1/2pound ground beef, browned
1 envelope taco seasoning
1 (10-ounce) can tomatoes with chilies, diced and drained

1 bell pepper, seeded and chopped
32 wonton wrappers
1 cup Cheddar cheese, shredded

Instructions:

Combine the pork, beef, taco seasoning, diced tomatoes, and bell pepper; mix well. Line all the muffin cups with wonton wrappers. Spritz with a nonstick cooking oil. Divide the beef filling among wrappers; top with the shredded cheese.

Cajun Sweet-Sour Grilled Pork

PREP: 5 MINUTES • COOK TIME: 12 MINUTES • TOTAL: 17 MINUTES • SERVES: 3

Ingredients

¼ cup brown sugar

1/4 cup cider vinegar

1-lb pork loin, sliced into 1-inch cubes

2 tablespoons Cajun seasoning

3 tablespoons brown sugar

Instructions:

1 In a shallow dish, mix well pork loin, 3 tablespoons brown sugar, and Cajun seasoning. Toss well to coat. Marinate in the ref for 3 hours. In a medium bowl mix well, brown sugar and vinegar for basting. Thread pork pieces in skewers. Baste with sauce and place on skewer rack in air fryer.

2 For 12 minutes, cook on 360°F. Halfway through cooking time, turnover skewers and baste with sauce. If needed, cook in batches. Serve and enjoy.

Chinese Braised Pork Belly

PREP: 5 MINUTES • COOK TIME: 20 MINUTES • TOTAL: 25 MINUTES • SERVES: 8

Ingredients

1 lb Pork Belly, sliced

1 Tbsp Oyster Sauce

1 Tbsp Sugar

2 Red Fermented Bean Curds

1 Tbsp Red Fermented Bean Curd Paste

1 Tbsp Cooking Wine

1/2 Tbsp Soy Sauce

1 Tsp Sesame Oil

1 Cup All Purpose Flour

Instructions:

1 Preheat the Innsky air fryer to 390 degrees. In a small bowl, mix all ingredients together and rub the pork thoroughly with this mixture. Set aside to marinate for at least 30 minutes or preferably overnight for the flavors to permeate the meat. Coat each marinated pork belly slice in flour and place in the Air fryer tray

2 Cook for 15 to 20 minutes until crispy and tender.

Air Fryer Sweet and Sour Pork

PREP: 10 MINUTES • COOK TIME: 12 MINUTES • TOTAL: 22 MINUTES • SERVES: 6

Ingredients

3 tbsp. olive oil

1/16 tsp. Chinese Five Spice

¼ tsp. pepper

½ tsp. sea salt

1 tsp. pure sesame oil

2 eggs

1 C. almond flour

2 pounds pork, sliced into chunks

Sweet and Sour Sauce:

¼ tsp. sea salt

½ tsp. garlic powder

1 tbsp. low-sodium soy sauce

½ C. rice vinegar

5 tbsp. tomato paste

1/8 tsp. water

½ C. sweetener of choice

Instructions:

1 To make the dipping sauce, whisk all sauce ingredients together over medium heat, stirring 5 minutes. Simmer uncovered 5 minutes till thickened. Meanwhile, combine almond flour, five spice, pepper, and salt. In another bowl, mix eggs with sesame oil. Dredge pork in flour mixture and then in egg mixture. Shake any excess off before adding to air fryer basket.

2 Set temperature to 340°F, and set time to 12 minutes. Serve with sweet and sour dipping sauce.

Per Serving: Calories: 371; Fat: 17g; Protein:27g; Sugar:1g

Pork Loin with Potatoes

PREP: 10 MINUTES • COOK TIME: 25 MINUTES • TOTAL: 35 MINUTES • SERVES: 2

Ingredients

2 pounds pork loin
1 teaspoon fresh parsley, chopped
2 large red potatoes, chopped

½ teaspoon garlic powder
½ teaspoon red pepper flakes, crushed
Salt and freshly ground black pepper, to taste

Instructions:

1 In a large bowl, add all ingredients except glaze and toss to coat well. Preheat the Innsky air fryer to 325 degrees F. Place the loin in the Air fryer basket. Arrange the potatoes around pork loin.

2 Cook for about 25 minutes.

Roasted Char Siew (Pork Butt)

PREP: 10 MINUTES • COOK TIME: 25 MINUTES • TOTAL: 35 MINUTES • SERVES: 6

Ingredients

1 strip of pork shoulder butt with a good amount of fat marbling
Marinade:
1 tsp. sesame oil
4 tbsp. raw honey

1 tsp. low-sodium dark soy sauce
1 tsp. light soy sauce
1 tbsp. rose wine
2 tbsp. Hoisin sauce

Instructions:

1 Combine all marinade ingredients together and add to Ziploc bag. Place pork in bag, making sure all sections of pork strip are engulfed in the marinade. Chill 3-24 hours.
Take out the strip 30 minutes before planning to cook and preheat your air fryer to 350 degrees.
Place foil on small pan and brush with olive oil. Place marinated pork strip onto prepared pan.

2 Set temperature to 350°F, and set time to 20 minutes. Roast 20 minutes. Glaze with marinade every 5-10 minutes. Remove strip and leave to cool a few minutes before slicing.

Asian Pork Chops

PREP: 2 HOURS 10 MINUTES • COOK TIME: 15 MINUTES • TOTAL:2 HOURS, 25 MINUTES • SERVES: 4

Ingredients

1/2 cup hoisin sauce
3 tablespoons cider vinegar
1 tablespoon Asian sweet chili sauce
1/4 teaspoon garlic powder

4 (1/2-inch-thick) boneless pork chops
1 teaspoon salt
1/2 teaspoon pepper

Instructions:

1. Stir together hoisin, chili sauce, garlic powder, and vinegar in a large mixing bowl. Separate 1/4 cup of this mixture, then add pork chops to the bowl and marinate in the fridge for 2 hours. Remove the pork chops and place them on a plate. Sprinkle each side of the pork chop evenly with salt and pepper.

2. Cook at 360 degrees for 14 minutes, flipping half way through. Brush with reserved marinade and serve.

Per Serving: Calories: 338; Fat: 21g; Protein:19g; Fiber:1g

Fried Pork Scotch Egg
PREP: 10 MINUTES • COOK TIME: 25 MINUTES • TOTAL: 35 MINUTES • SERVES: 2

Ingredients
3 soft-boiled eggs, peeled

8 ounces of raw minced pork, or sausage outside the casings

2 teaspoons of ground rosemary

2 teaspoons of garlic powder

Pinch of salt and pepper

2 raw eggs

1 cup of breadcrumbs

Instructions:
1 Cover the basket of the Air fryer with a lining of tin foil, leaving the edges uncovered to allow air to circulate through the basket. Preheat the Innsky air fryer to 350 degrees. In a mixing bowl, combine the raw pork with the rosemary, garlic powder, salt and pepper. This will probably be easiest to do with your masher or bare hands, combine until all the spices are evenly spread throughout the meat. Divide the meat mixture into three equal portions in the mixing bowl, and form each into balls with your hands. Lay a large sheet of plastic wrap on the countertop, and flatten one of the balls of meat on top of it, to form a wide, flat meat-circle. Place one of the peeled soft-boiled eggs in the center of the meat-circle and then, using the ends of the plastic wrap, pull the meat-circle so that it is fully covering and surrounding the soft-boiled egg. Tighten and shape the plastic wrap covering the meat so that if forms a ball, and make sure not to squeeze too hard lest you squish the soft-boiled egg at the center of the ball! Set aside. Repeat steps 5-7 with the other two soft-boiled eggs and portions of meat-mixture. In a separate mixing bowl, beat the two raw eggs until fluffy and until the yolks and whites are fully combined. One by one, remove the plastic wrap and dunk the pork-covered balls into the raw egg, and then roll them in the bread crumbs, covering fully and generously. Place each of the bread-crumb covered meat-wrapped balls onto the foil-lined surface of the air fryer. Three of them should fit nicely, without touching.

2 Set the Innsky air fryer timer to 25 minutes. About halfway through the cooking time, shake the handle of the air-fryer vigorously, so that the scotch eggs inside roll around and ensure full coverage. After 25 minutes, the air fryer will shut off and the scotch eggs should be perfect – the meat fully cooked, the egg-yolks still runny on the inside, and the outsides crispy and golden-brown. Using tongs, place them on serving plates, slice in half, and enjoy.

Juicy Pork Ribs Ole
PREP: 10 MINUTES • COOK TIME: 25 MINUTES • TOTAL: 35 MINUTES • SERVES: 4

Ingredients
1 rack of pork ribs

1/2 cup low-fat milk

1 tablespoon envelope taco seasoning mix

1 can tomato sauce

1/2 teaspoon ground black pepper

1 teaspoon seasoned salt

1 tablespoon cornstarch

1 teaspoon canola oil

Instructions:
1. Place all ingredients in a mixing dish; let them marinate for 1 hour.
2. Cook the marinated ribs approximately 25 minutes at 390 degrees F
    Work with batches. Enjoy

Vietnamese Pork Chops
PREP: 10 MINUTES • COOK TIME: 7 MINUTES • TOTAL: 25 MINUTES • SERVES: 6

Ingredients
1 tbsp. olive oil
1 tbsp. fish sauce
1 tsp. low-sodium dark soy sauce
1 tsp. pepper
3 tbsp. lemongrass

1 tbsp. chopped shallot
1 tbsp. chopped garlic
1 tbsp. brown sugar
2 pork chops

Instructions:
1   Add pork chops to a bowl along with olive oil, fish sauce, soy sauce, pepper, lemongrass, shallot, garlic, and brown sugar.
Marinade pork chops 2 hours.
Ensure your air fryer is preheated to 400 degrees. Add pork chops to the basket.
2   Set temperature to 400°F, and set time to 7 minutes. Cook making sure to flip after 5 minutes of cooking.
Serve alongside steamed cauliflower rice.
Per Serving: Calories: 290; Fat: 15g; Protein:30g; Sugar:3g

Ham and Cheese Rollups
PREP: 5 MINUTES • COOK TIME: 8 MINUTES • TOTAL: 15 MINUTES • SERVES: 12

Ingredients
2 tsp. raw honey
2 tsp. dried parsley
1 tbsp. poppy seeds
½ C. melted coconut oil

¼ C. spicy brown mustard
9 slices of provolone cheese
10 ounces of thinly sliced Black Forest Ham
1 tube of crescent rolls

Instructions:
1   Roll out dough into a rectangle. Spread 2-3 tablespoons of spicy mustard onto dough, then layer provolone cheese and ham slices.
Roll the filled dough up as tight as you can and slice into 12-15 pieces.
Melt coconut oil and mix with a pinch of salt and pepper, parsley, honey, and remaining mustard. Brush mustard mixture over roll-ups and sprinkle with poppy seeds.
Grease Innsky air fryer basket liberally with olive oil and add rollups.
2   Set temperature to 350°F, and set time to 8 minutes.
Serve!
Per serving: calories: 289; fat: 6g; protein:18Beef Recipes

Beef recipes

Cheeseburger Egg Rolls
PREP: 10 MINUTES • COOK TIME: 7 MINUTES • TOTAL: 17 MINUTES • SERVES: 6

Ingredients

6 egg roll wrappers
6 chopped dill pickle chips
1 tbsp. yellow mustard
3 tbsp. cream cheese
3 tbsp. shredded cheddar cheese

½ C. chopped onion
½ C. chopped bell pepper
¼ tsp. onion powder
¼ tsp. garlic powder
8 ounces of raw lean ground beef

Instructions:

1   In a skillet, add seasonings, beef, onion, and bell pepper. Stir and crumble beef till fully cooked, and vegetables are soft. Take skillet off the heat and add cream cheese, mustard, and cheddar cheese, stirring till melted. Pour beef mixture into a bowl and fold in pickles.

Lay out egg wrappers and place 1/6th of beef mixture into each one. Moisten egg roll wrapper edges with water. Fold sides to the middle and seal with water. Repeat with all other egg rolls.

Place rolls into air fryer, one batch at a time.

2   Set temperature to 392°F, and set time to 7 minutes.

Per Serving: Calories: 153; Fat: 4g; Protein:12g; Sugar:3g

~~Air Fried Grilled Steak~~
PREP: 5 MINUTES • COOK TIME: 45 MINUTES • TOTAL: 50 MINUTES • SERVES: 2

Ingredients

2 top sirloin steaks
3 tablespoons butter, melted

3 tablespoons olive oil
Salt and pepper to taste

Instructions:

1   Preheat the Innsky air fryer for 5 minutes. Season the sirloin steaks with olive oil, salt and pepper. Place the beef in the air fryer basket.

2   Cook for ~~45~~ minutes at ~~350°F~~. Once cooked, serve with butter.

Preheat 400°F.
med - Rare — 10 mins
medium — 12 mins.

Per Serving: Calories: 1536; Fat: 123.7g; Protein:103.4g

Beef Brisket Recipe from Texas *med. well — 14 mins*
PREP: 15 MINUTES • COOK TIME: 1HOUR AND 30 MINUTES • SERVES: 8

Ingredients

1 ½ cup beef stock
1 bay leaf
1 tablespoon garlic powder
1 tablespoon onion powder
2 pounds beef brisket, trimmed

2 tablespoons chili powder
2 teaspoons dry mustard
4 tablespoons olive oil
Salt and pepper to taste

Instructions:

1   Preheat the Innsky air fryer for 5 minutes. Place all ingredients in a deep baking dish that will fit in the air fryer.

2   Bake for 1 hour and 30 minutes at 400°F.

Stir the beef every after 30 minutes to soak in the sauce.

Per Serving: Calories: 306; Fat: 24.1g; Protein:18.3g

Savory Beefy Poppers
PREP: 15 MINUTES • COOK TIME: 15 MINUTES • TOTAL: 30 MINUTES • SERVES: 8

Ingredients

8 medium jalapeño peppers, stemmed, halved, and seeded

1 (8-ounce) package cream cheese softened

2 pounds ground beef (85% lean)

1 teaspoon fine sea salt

½ teaspoon ground black pepper

8 slices thin-cut bacon

Fresh cilantro leaves, for garnish

Instructions:

1. Spray the Innsky air fryer basket with avocado oil. Preheat the Innsky air fryer to 400°F. Stuff each jalapeño half with a few tablespoons of cream cheese. Place the halves back together again to form 8 jalapeños. Season the ground beef with the salt and pepper and mix with your hands to incorporate. Flatten about ¼ pound of ground beef in the palm of your hand and place a stuffed jalapeño in the center. Fold the beef around the jalapeño, forming an egg shape. Wrap the beef-covered jalapeño with a slice of bacon and secure it with a toothpick.

2. Place the jalapeños in the air fryer basket, leaving space between them (if you're using a smaller air fryer, work in batches if necessary), and cook for 15 minutes, or until the beef is cooked through and the bacon is crispy. Garnish with cilantro before serving. Store leftovers in an airtight container in the fridge for 3 days or in the freezer for up to a month. Reheat in a preheated 350°F air fryer for 4 minutes, or until heated through and the bacon is crispy.

Per serving: Calories 679; Fat 53g; Protein 42g; Total carbs 3g; Fiber 1g

Juicy Cheeseburgers
PREP: 5 MINUTES • COOK TIME: 15 MINUTES • TOTAL: 20 MINUTES • SERVES: 4

Ingredients

1 pound 93% lean ground beef

1 teaspoon Worcestershire sauce

1 tablespoon burger seasoning

Salt

Pepper

Cooking oil

4 slices cheese

2 buns

Instructions:

1 In a large bowl, mix the ground beef, Worcestershire, burger seasoning, and salt and pepper to taste until well blended. Spray the Innsky air fryer basket with cooking oil. You will need only a quick spritz. The burgers will produce oil as they cook. Shape the mixture into 4 patties. Place the burgers in the air fryer. The burgers should fit without the need to stack, but stacking is okay if necessary.

2 Cook for 8 minutes. Open the air fryer and flip the burgers. Cook for an additional 3 to 4 minutes. Check the inside of the burgers to determine if they have finished cooking. You can stick a knife or fork in the center to examine the color. Top each burger with a slice of cheese. Cook for an additional minute, or until the cheese has melted. Serve on buns with any additional toppings of your choice.

Per Serving: Calories: 566; Fat: 39g; Protein:29g; Fiber:1g

Copycat Taco Bell Crunch Wraps

PREP: 10 MINUTES • COOK TIME: 2 MINUTES • TOTAL: 15 MINUTES • SERVES: 6

Ingredients

6 wheat tostadas
2 C. sour cream
2 C. Mexican blend cheese
2 C. shredded lettuce
12 ounces low-sodium nacho cheese

3 Roma tomatoes
6 12-inch wheat tortillas
1 1/3 C. water
2 packets low-sodium taco seasoning
2 pounds of lean ground beef

Instructions:

1 Ensure your air fryer is preheated to 400 degrees.
   Make beef according to taco seasoning packets.
   Place 2/3 C. prepared beef, 4 tbsp. cheese, 1 tostada, 1/3 C. sour cream, 1/3 C. lettuce, 1/6th of tomatoes and 1/3 C. cheese on each tortilla.
   Fold up tortillas edges and repeat with remaining ingredients.
   Lay the folded sides of tortillas down into the air fryer and spray with olive oil.
2 Set temperature to 400°F, and set time to 2 minutes. Cook 2 minutes till browned.

Per Serving: Calories: 311; Fat: 9g; Protein:22g; Sugar:2g

Swedish Meatloaf

PREP: 10 MINUTES • COOK TIME: 35 MINUTES • TOTAL: 45 MINUTES • SERVES: 8

Ingredients

1½ pounds ground beef (85% lean)
¼ pound ground pork or ground beef
1 large egg (omit for egg-free)
½ cup minced onions
¼ cup tomato sauce
2 tablespoons dry mustard
2 cloves garlic, minced
2 teaspoons fine sea salt
1 teaspoon ground black pepper, plus more for garnish

SAUCE:
½ cup (1 stick) unsalted butter
½ cup shredded Swiss or mild cheddar cheese (about 2 ounces)
2 ounces cream cheese (¼ cup), softened
⅓ cup beef broth
⅛ teaspoon ground nutmeg
Halved cherry tomatoes, for serving (optional)

Instructions:

1. Preheat the Innsky air fryer to 390°F.

    In a large bowl, combine the ground beef, ground pork, egg, onions, tomato sauce, dry mustard, garlic, salt, and pepper. Using your hands, mix until well combined.
2. Place the meatloaf mixture in a 9 by 5-inch loaf pan and place it in the air fryer. Cook for 35 minutes, or until cooked through and the internal temperature reaches 145°F. Check the meatloaf after 25 minutes; if it's getting too brown on the top, cover it loosely with foil to prevent burning.

    While the meatloaf cooks, make the sauce: Heat the butter in a saucepan over medium-high heat until it sizzles and brown flecks appear, stirring constantly to keep the butter from burning. Turn the heat down to low and whisk in the Swiss cheese, cream cheese, broth, and nutmeg. Simmer for at least 10 minutes. The longer it simmers, the more the flavors open up. When the meatloaf is done, transfer it to a serving tray and pour the sauce over it. Garnish with ground black pepper and serve with cherry tomatoes, if desired. Allow the meatloaf to rest for 10 minutes before slicing so it doesn't crumble apart.

    Store leftovers in an airtight container in the fridge for 3 days or in the freezer for up to a month. Reheat in a preheated 350°F air fryer for 4 minutes, or until heated through.

Per serving: Calories 395; Fat 32g; Protein 23g; Total carbs 3g; Fiber 1g

Carne Asada

PREP: 5 MINUTES PLUS 2 HOURS TO MARINATE•COOK TIME: 8 MINUTES•TOTAL:2 HOURS 13 MINUTES•SERVES: 4

Ingredients

MARINADE:

1 cup fresh cilantro leaves and stems, plus more for garnish if desired

1 jalapeño pepper, seeded and diced

½ cup lime juice

2 tablespoons avocado oil

2 tablespoons coconut vinegar or apple cider vinegar

2 teaspoons orange extract

1 teaspoon stevia glycerite, or ⅛ teaspoon liquid stevia

2 teaspoons ancho chili powder

2 teaspoons fine sea salt

1 teaspoon coriander seeds

1 teaspoon cumin seeds

1pound skirt steak, cut into 4 equal portions

FOR SERVING (OPTIONAL):

Chopped avocado

Lime slices

Sliced radishes

Instructions:

1. Make the marinade: Place all the ingredients for the marinade in a blender and puree until smooth. Place the steak in a shallow dish and pour the marinade over it, making sure the meat is covered completely. Cover and place in the fridge for 2 hours or overnight.

   Spray the Innsky air fryer basket with avocado oil. Preheat the Innsky air fryer to 400°F.

2. Remove the steak from the marinade and place it in the Innsky air fryer basket in one layer. Cook for 8 minutes, or until the internal temperature is 145°F; do not overcook or it will become tough.

   Remove the steak from the air fryer and place it on a cutting board to rest for 10 minutes before slicing it against the grain. Garnish with cilantro, if desired, and serve with chopped avocado, lime slices, and/or sliced radishes, if desired. Store leftovers in an airtight container in the fridge for 3 days or in the freezer for up to a month. Reheat in a preheated 350°F air fryer for 4 minutes, or until heated through.

Per serving: Calories 263; Fat 17g; Protein 24g; Total carbs 4g; Fiber 1g

Spicy Thai Beef Stir-Fry

PREP: 15 MINUTES • COOK TIME: 9 MINUTES • TOTAL: 24 MINUTES • SERVES: 4

Ingredients

1pound sirloin steaks, thinly sliced

2 tablespoons lime juice, divided

⅓ cup crunchy peanut butter

½ cup beef broth

1 tablespoon olive oil

1½ cups broccoli florets

2 cloves garlic, sliced

1 to 2 red chile peppers, sliced

Instructions:

1. In a medium bowl, combine the steak with 1 tablespoon of the lime juice. Set aside.

   Combine the peanut butter and beef broth in a small bowl and mix well. Drain the beef and add the juice from the bowl into the peanut butter mixture.

   In a 6-inch metal bowl, combine the olive oil, steak, and broccoli.

2. Cook for 3 to 4 minutes or until the steak is almost cooked and the broccoli is crisp and tender, shaking the basket once during cooking time.

   Add the garlic, chile peppers, and the peanut butter mixture and stir.

   Cook for 3 to 5 minutes or until the sauce is bubbling and the broccoli is tender.

   Serve over hot rice.

Per Serving: Calories: 387; Fat: 22g; Protein:42g; Fiber:2g

Air Fryer Beef Casserole
PREP: 5 MINUTES • COOK TIME: 30 MINUTES • TOTAL: 35 MINUTES • SERVES: 4

Ingredients

1 green bell pepper, seeded and chopped
1 onion, chopped
1-pound ground beef
3 cloves of garlic, minced

3 tablespoons olive oil
6 cups eggs, beaten
Salt and pepper to taste

Instructions:

1   Preheat the Innsky air fryer for 5 minutes. In a baking dish that will fit in the air fryer, mix the ground beef, onion, garlic, olive oil, and bell pepper. Season with salt and pepper to taste.
Pour in the beaten eggs and give a good stir. Place the dish with the beef and egg mixture in the air fryer.
2   Bake for 30 minutes at 325°F.

Per Serving: Calories: 1520; Fat: 125.11g; Protein:87.9g

Salisbury Steak with Mushroom Onion Gravy
PREP: 10 MINUTES • COOK TIME: 33 MINUTES • TOTAL: 43 MINUTES • SERVES: 2

Ingredients

MUSHROOM ONION GRAVY:
¾ cup sliced button mushrooms
¼ cup thinly sliced onions
¼ cup unsalted butter, melted (or bacon fat for dairy-free)
½ teaspoon fine sea salt
¼ cup beef broth
STEAKS:
½ pound ground beef (85% lean)

¼ cup minced onions, or ½ teaspoon onion powder
2 tablespoons tomato paste
1 tablespoon dry mustard
1 clove garlic, minced, or ¼ teaspoon garlic powder
½ teaspoon fine sea salt
¼ teaspoon ground black pepper, plus more for garnish if desired
Chopped fresh thyme leaves, for garnish (optional)

Instructions:

1. Preheat the Innsky air fryer to 390°F.
Make the gravy: Place the mushrooms and onions in a casserole dish that will fit in your air fryer. Pour the melted butter over them and stir to coat, then season with the salt. Place the dish in the air fryer and cook for 5 minutes, stir, then cook for another 3 minutes, or until the onions are soft and the mushrooms are browning. Add the broth and cook for another 10 minutes.
While the gravy is cooking, prepare the steaks: In a large bowl, mix together the ground beef, onions, tomato paste, dry mustard, garlic, salt, and pepper until well combined. Form the mixture into 2 oval-shaped patties.
2. Place the patties on top of the mushroom gravy. Cook for 10 minutes, gently flip the patties, then cook for another 2 to 5 minutes, until the beef is cooked through and the internal temperature reaches 145°F.
Transfer the steaks to a serving platter and pour the gravy over them. Garnish with ground black pepper and chopped fresh thyme, if desired. Store leftovers in an airtight container in the fridge for 3 days or in the freezer for up to a month. Reheat in a preheated 350°F air fryer for 4 minutes, or until heated through.

Per serving: Calories 588; Fat 44g; Protein 33g; Total carbs 11g; Fiber 3g

Fajita Meatball Lettuce Wraps

PREP: 10 MINUTES • COOK TIME: 10 MINUTES • TOTAL: 20 MINUTES • SERVES: 4

Ingredients

1pound ground beef (85% lean)
½ cup salsa, plus more for serving if desired
¼ cup chopped onions
¼ cup diced green or red bell peppers
1 large egg, beaten
1 teaspoon fine sea salt
½ teaspoon chili powder

½ teaspoon ground cumin
1 clove garlic, minced
FOR SERVING (OPTIONAL):
8 leaves Boston lettuce
Pico de gallo or salsa
Lime slices

Instructions:

1. Spray the Innsky air fryer basket with avocado oil. Preheat the Innsky air fryer to 350°F. In a large bowl, mix together all the ingredients until well combined.
   Shape the meat mixture into eight 1-inch balls.
2. Place the meatballs in the air fryer basket, leaving a little space between them. Cook for 10 minutes, or until cooked through and no longer pink inside and the internal temperature reaches 145°F. Serve each meatball on a lettuce leaf, topped with pico de gallo or salsa, if desired. Serve with lime slices if desired. Store leftovers in an airtight container in the fridge for 3 days or in the freezer for up to a month. Reheat in a preheated 350°F air fryer for 4 minutes, or until heated through.

Per serving: Calories 272; Fat 18g; Protein 23g; Total carbs 3g; Fiber 0.5g

Chimichurri Skirt Steak

PREP: 10 MINUTES • COOK TIME: 8 MINUTES • TOTAL: 18 MINUTES • SERVES: 2

Ingredients

2 x 8 oz Skirt Steak
1 cup Finely Chopped Parsley
¼ cup Finely Chopped Mint
2 Tbsp Fresh Oregano (Washed & finely chopped)
3 Finely Chopped Cloves of Garlic
1 Tsp Red Pepper Flakes (Crushed)
1 Tbsp Ground Cumin

1 Tsp Cayenne Pepper
2 Tsp Smoked Paprika
1 Tsp Salt
¼ Tsp Pepper
¾ cup Oil
3 Tbsp Red Wine Vinegar

Instructions:

1. Throw all the ingredients in a bowl (besides the steak) and mix well.
   Put ¼ cup of the mixture in a plastic baggie with the steak and leave in the fridge overnight (2–24hrs).
2. Leave the bag out at room temperature for at least 30 min before popping into the Air fryer. Preheat for a minute or two to 390° F before cooking until med–rare (8–10 min).
   Put 2 Tbsp of the chimichurri mix on top of each steak before serving.

Reuben Fritters

PREP: 10 MINUTES • COOK TIME: 16 MINUTES • TOTAL: 17 MINUTES • SERVES: 1 dozen fritters

Ingredients

2 cups finely diced cooked corned beef
1 (8-ounce) package cream cheese, softened
½ cup finely shredded Swiss cheese (about 2 ounces)
¼ cup sauerkraut

1 cup pork dust or powdered Parmesan cheese
Chopped fresh thyme, for garnish
Thousand Island Dipping Sauce for serving
Cornichons, for serving (optional)

Instructions:

1. Spray the Innsky air fryer basket with avocado oil. Preheat the Innsky air fryer to 390°F. In a large bowl, mix together the corned beef, cream cheese, Swiss cheese, and sauerkraut until well combined. Form the corned beef mixture into twelve 1½-inch balls.

   Place the pork dust in a shallow bowl. Roll the corned beef balls in the pork dust and use your hands to form it into a thick crust around each ball.

2. Place 6 balls in the air fryer basket, spaced about ½ inch apart, and cook for 8 minutes, or until golden brown and crispy. Allow them to cool a bit before lifting them out of the air fryer (the fritters are very soft when the cheese is melted; they're easier to handle once the cheese has hardened a bit). Repeat with the remaining fritters.

   Garnish with chopped fresh thyme and serve with the dipping sauce and cornichons, if desired. Store leftovers in an airtight container in the fridge for 3 days or in the freezer for up to a month. Reheat in a preheated 350°F air fryer for 4 minutes, or until heated through.

Per serving: Calories 527; Fat 50g; Protein 18g; Total carbs 2g; Fiber 0.1g

Charred Onions And Steak Cube BBQ

PREP: 5 MINUTES • COOK TIME: 40 MINUTES • TOTAL: 45 MINUTES • SERVES: 3

Ingredients

1 cup red onions, cut into wedges
1 tablespoon dry mustard
1 tablespoon olive oil

1-pound boneless beef sirloin, cut into cubes
Salt and pepper to taste

Instructions:

1 Preheat the Innsky air fryer to 390°F.

   Place the grill pan accessory in the air fryer.

   Toss all ingredients in a bowl and mix until everything is coated with the seasonings.

2 Place on the grill pan and cook for 40 minutes.

   Halfway through the cooking time, give a stir to cook evenly.

Per Serving: Calories: 260; Fat: 10.7g; Protein:35.5g

Steak and Mushroom Gravy

PREP: 15 MINUTES • COOK TIME: 15 MINUTES • TOTAL: 30 MINUTES • SERVES: 4

Ingredients

4 cubed steaks
2 large eggs
1/2 dozen mushrooms
4 tablespoons unsalted butter
4 tablespoons black pepper
2 tablespoons salt
1/2 teaspoon onion powder

1/2 teaspoon garlic powder
1/4 teaspoon cayenne powder
1 1/4 teaspoons paprika
1 1/2 cups whole milk
1/3 cup flour
2 tablespoons vegetable oil

Instructions:

1   Mix 1/2 flour and a pinch of black pepper in a shallow bowl or on a plate. Beat 2 eggs in a bowl and mix in a pinch of salt and pepper.
In another shallow bowl mix together the other half of the flour with a pepper to taste, garlic powder, paprika, cayenne, and onion powder.
Chop mushrooms and set aside. Press your steak into the first flour bowl, then dip in egg, then press the steak into the second flour bowl until covered completely.

2   Cook steak in your Air fryer at 360 degrees for 15 Minutes, flipping halfway through. While the steak cooks, warm the butter over medium heat and add mushrooms to sauté. Add 4 tablespoons of the flour and pepper mix to the pan and mix until there are no clumps of flour. Mix in whole milk and simmer. Serve over steak for breakfast, lunch, or dinner.

Per Serving: Calories: 442; Fat: 27g; Protein:32g; Fiber:2.3g

Country Fried Steak

PREP: 5 MINUTES • COOK TIME: 12 MINUTES • TOTAL: 20 MINUTES • SERVES: 2

Ingredients

1 tsp. pepper
2 C. almond milk
2 tbsp. almond flour
6 ounces ground sausage meat
1 tsp. pepper
1 tsp. salt

1 tsp. garlic powder
1 tsp. onion powder
1 C. panko breadcrumbs
1 C. almond flour
3 beaten eggs
6 ounces sirloin steak, pounded till thin

Instructions:

1   Season panko breadcrumbs with spices. Dredge steak in flour, then egg, and then seasoned panko mixture. Place into air fryer basket.

2   Set temperature to 370°F, and set time to 12 minutes. To make sausage gravy, cook sausage and drain off fat, but reserve 2 tablespoons. Add flour to sausage and mix until incorporated. Gradually mix in milk over medium to high heat till it becomes thick. Season mixture with pepper and cook 3 minutes longer. Serve steak topped with gravy and enjoy.

Per Serving: Calories: 395; Fat: 11g; Protein:39g; Sugar:5g

Greek Stuffed Tenderloin
PREP: 10 MINUTES • COOK TIME: 10 MINUTES • TOTAL: 20 MINUTES • SERVES: 4

Ingredients

1½ pounds venison or beef tenderloin, pounded to ¼ inch thick
3 teaspoons fine sea salt
1 teaspoon ground black pepper
2 ounces creamy goat cheese
½ cup crumbled feta cheese (about 2 ounces)
¼ cup finely chopped onions

2 cloves garlic, minced
FOR GARNISH/SERVING (OPTIONAL):
Prepared yellow mustard
Halved cherry tomatoes
Extra-virgin olive oil
Sprigs of fresh rosemary
Lavender flowers

Instructions:

1. Spray the Innsky air fryer basket with avocado oil. Preheat the Innsky air fryer to 400°F. Season the tenderloin on all sides with the salt and pepper.

   In a medium-sized mixing bowl, combine the goat cheese, feta, onions, and garlic. Place the mixture in the center of the tenderloin. Starting at the end closest to you, tightly roll the tenderloin like a jelly roll. Tie the rolled tenderloin tightly with kitchen twine.

2. Place the meat in the Innsky air fryer basket and cook for 5 minutes. Flip the meat over and cook for another 5 minutes, or until the internal temperature reaches 135°F for medium-rare. To serve, smear a line of prepared yellow mustard on a platter, then place the meat next to it and add halved cherry tomatoes on the side, if desired. Drizzle with olive oil and garnish with rosemary sprigs and lavender flowers, if desired.

   Best served fresh. Store leftovers in an airtight container in the fridge for 3 days. Reheat in a preheated 350°F air fryer for 4 minutes, or until heated through.

Per serving: Calories **415**; Fat **16g**; Protein **62g**; Total carbs **4g**; Fiber **0.3g**

Warming Winter Beef with Celery
PREP: 5 MINUTES • COOK TIME: 12 MINUTES • TOTAL: 15 MINUTES • SERVES: 4

Ingredients

9 ounces tender beef, chopped
1/2 cup leeks, chopped
1/2 cup celery stalks, chopped
2 cloves garlic, smashed
2 tablespoons red cooking wine

3/4 cup cream of celery soup
2 sprigs rosemary, chopped
1/4 teaspoon smoked paprika
3/4 teaspoons salt
1/4 teaspoon black pepper, or to taste

Instructions:

1 Add the beef, leeks, celery, and garlic to the baking dish; cook for about 5 minutes at 390 degrees F.

   Once the meat is starting to tender, pour in the wine and soup. Season with rosemary, smoked paprika, salt, and black pepper. Now, cook an additional 7 minutes.

Black and Blue Burgers

PREP: 5 MINUTES • COOK TIME:10 MINUTES • TOTAL: 15 MINUTES • SERVES: 2

Ingredients

½ teaspoon fine sea salt

¼ teaspoon ground black pepper

¼ teaspoon garlic powder

¼ teaspoon onion powder

¼ teaspoon smoked paprika

2 (¼-pound) hamburger patties, ½ inch thick

½ cup crumbled blue cheese (about 2 ounces)

2 Hamburger Buns

2 tablespoons mayonnaise

6 red onion slices

2 Boston lettuce leaves

Instructions:

1   Spray the Innsky air fryer basket with avocado oil. Preheat the Innsky air fryer to 360°F.In a small bowl, combine the salt, pepper, and seasonings. Season the patties well on both sides with the seasoning mixture.

2   Place the patties in the Innsky air fryer basket and cook for 7 minutes, or until the internal temperature reaches 145°F for a medium-done burger. Place the blue cheese on top of the patties and cook for another minute to melt the cheese. Remove the burgers from the air fryer and allow to rest for 5 minutes. Slice the buns in half and smear 2 halves with a tablespoon of mayo each. Increase the heat to 400°F and place the buns in the Innsky air fryer basket cut side up. Toast the buns for 1 to 2 minutes, until golden brown. Remove the buns from the air fryer and place them on a serving plate. Place the burgers on the buns and top each burger with 3 red onion slices and a lettuce leaf.

Per serving: Calories 237; Fat 20g; Protein 11g; Total carbs 3g; Fiber 1g

Cheesy Ground Beef And Mac Taco Casserole

PREP: 10 MINUTES • COOK TIME: 25 MINUTES • TOTAL: 35 MINUTES • SERVES: 5

Ingredients

1-ounce shredded Cheddar cheese

1-ounce shredded Monterey Jack cheese

2 tablespoons chopped green onions

1/2 (10.75 ounce) can condensed tomato soup

1/2-pound lean ground beef

1/2 cup crushed tortilla chips

1/4-pound macaroni, cooked according to manufacturer's Instructions

1/4 cup chopped onion

1/4 cup sour cream (optional)

1/2 (1.25 ounce) package taco seasoning mix

1/2 (14.5 ounce) can diced tomatoes

Instructions:

1   Lightly grease baking pan of air fryer with cooking spray. Add onion and ground beef. For 10 minutes, cook on 360°F. Halfway through cooking time, stir and crumble ground beef. Add taco seasoning, diced tomatoes, and tomato soup. Mix well. Mix in pasta. Sprinkle crushed tortilla chips. Sprinkle cheese.

2   Cook for 15 minutes at 390°F until tops are lightly browned and cheese is melted. Serve and enjoy.

Meat Lovers' Pizza

PREP: 10 MINUTES • COOK TIME: 12 MINUTES • TOTAL: 22 MINUTES • SERVES: 2

Ingredients

1 pre-prepared 7-inch pizza pie crust, defrosted if necessary.

1/3 cup of marinara sauce.

2 ounces of grilled steak, sliced into bite-sized pieces

2 ounces of salami, sliced fine

2 ounces of pepperoni, sliced fine

¼ cup of American cheese

¼ cup of shredded mozzarella cheese

Instructions:

1   Preheat the Innsky air fryer to 350 degrees. Lay the pizza dough flat on a sheet of parchment paper or tin foil, cut large enough to hold the entire pie crust, but small enough that it will leave the edges of the air frying basket uncovered to allow for air circulation. Using a fork, stab the pizza dough several times across the surface – piercing the pie crust will allow air to circulate throughout the crust and ensure even cooking. With a deep soup spoon, ladle the marinara sauce onto the pizza dough, and spread evenly in expanding circles over the surface of the pie-crust. Be sure to leave at least ½ inch of bare dough around the edges, to ensure that extra-crispy crunchy first bite of the crust! Distribute the pieces of steak and the slices of salami and pepperoni evenly over the sauce-covered dough, then sprinkle the cheese in an even layer on top.

2   Set the Innsky air fryer timer to 12 minutes, and place the pizza with foil or paper on the fryer's basket surface. Again, be sure to leave the edges of the basket uncovered to allow for proper air circulation, and don't let your bare fingers touch the hot surface. After 12 minutes, when the Air fryer shuts off, the cheese should be perfectly melted and lightly crisped, and the pie crust should be golden brown. Using a spatula – or two, if necessary, remove the pizza from the Innsky air fryer basket and set on a serving plate. Wait a few minutes until the pie is cool enough to handle, then cut into slices and serve.

Mushroom and Swiss Burgers

PREP: 5 MINUTES • COOK TIME: 15 MINUTES • TOTAL: 20 MINUTES • SERVES: 2

Ingredients

2 large portobello mushrooms

1 teaspoon fine sea salt, divided

¼ teaspoon garlic powder

¼ teaspoon ground black pepper

¼ teaspoon onion powder

¼ teaspoon smoked paprika

2 (¼-pound) hamburger patties, ½ inch thick

2 slices Swiss cheese (omit for dairy-free)

Condiments of choice, such as Ranch Dressing, prepared yellow mustard, or mayonnaise, for serving

Instructions:

1   Preheat the Innsky air fryer to 360°F. Clean the portobello mushrooms and remove the stems. Spray the mushrooms on all sides with avocado oil and season them with ½ teaspoon of the salt.

2   Place the mushrooms in the Innsky air fryer basket and cook for 7 to 8 minutes, until fork-tender and soft to the touch.

While the mushrooms cook, in a small bowl mix together the remaining ½ teaspoon of salt, the garlic powder, pepper, onion powder, and paprika. Sprinkle the hamburger patties with the seasoning mixture. When the mushrooms are done cooking, remove them from the air fryer and place them on a serving platter with the cap side down.

Place the hamburger patties in the air fryer and cook for 7 minutes, or until the internal temperature reaches 145°F for a medium-done burger. Place a slice of Swiss cheese on each patty and cook for another minute to melt the cheese. Place the burgers on top of the mushrooms and drizzle with condiments of your choice. Best served fresh.

Per serving: Calories 345; Fat 23g; Protein 30g; Total carbs 5g; Fiber 1g

Air Fryer Steak Tips

PREP: 5 MINUTES PLUS 1 HOUR TO MARINATE • COOK TIME: 8 MINUTES • TOTAL: 1 HOUR 13 MINUTES• SERVES: 4

Ingredients

⅓ cup soy sauce
1 cup water
¼ cup freshly squeezed lemon juice
3 tablespoons brown sugar

1 teaspoon garlic powder
1 teaspoon ground ginger
1 teaspoon dried parsley
2 pounds steak tips, cut into 1-inch cubes

Instructions:

1   In a large mixing bowl, make the marinade. Mix together the soy sauce, water, lemon juice, brown sugar, garlic powder, ginger, and parsley. Place the meat in the marinade, then cover and refrigerate for at least 1 hour. Preheat the Innsky air fryer to 400°F. Spray the Innsky air fryer basket with olive oil.
When the steak is done marinating, place it in the greased air fryer basket.

2   Set the timer and cook for 4 minutes. Using tongs, flip the meat. Reset the timer and cook for 4 minutes more.

Per Serving: Calories: 347; Fat: 9g; Saturated fat: 1g; Carbohydrate: 9g; Fiber: 0g; Sugar: 7g; Protein: 53g; Iron: 1mg; Sodium: 1203mg

Creamy Burger & Potato Bake

PREP: 5 MINUTES • COOK TIME: 55 MINUTES • TOTAL: 60 MINUTES • SERVES: 3

Ingredients

salt to taste
freshly ground pepper, to taste
1/2 (10.75 ounce) can condensed cream of mushroom soup
1/2-pound lean ground beef

1-1/2 cups peeled and thinly sliced potatoes
1/2 cup shredded Cheddar cheese
1/4 cup chopped onion
1/4 cup and 2 tablespoons milk

Instructions:

1   Lightly grease baking pan of air fryer with cooking spray. Add ground beef. For 10 minutes, cook on 360°F. Stir and crumble halfway through cooking time. Meanwhile, in a bowl, whisk well pepper, salt, milk, onion, and mushroom soup. Mix well. Drain fat off ground beef and transfer beef to a plate. In same air fryer baking pan, layer ½ of potatoes on bottom, then ½ of soup mixture, and then ½ of beef. Repeat process. Cover pan with foil.

2   Cook for 30 minutes. Remove foil and cook for another 15 minutes or until potatoes are tender. Serve and enjoy.

Per Serving: Calories: 399; Fat: 26.9g; Protein:22.1g

Beef Stroganoff

PREP: 10 MINUTES • COOK TIME: 14 MINUTES • TOTAL: 24 MINUTES • SERVES: 4

Ingredients

9 Ozs Tender Beef
1 Onion, chopped
1 Tbsp Paprika
3/4 Cup Sour Cream

Salt and Pepper to taste
Baking Dish

Instructions:

1.   Preheat the Innsky air fryer to 390 degrees.

2.   Chop the beef and marinate it with the paprika. Add the chopped onions into the baking dish and heat for about 2 minutes in the Air fryer. When the onions are transparent, add the beef into the dish and cook for 5 minutes. Once the beef is starting to tender, pour in the sour cream and cook for another 7 minutes. At this point, the liquid should have reduced. Season with salt and pepper and serve.

Air Fried Beef Stir-Fry

PREP: 10 MINUTES PLUS 30 MINUTES TO MARINATE • COOK TIME: 10 MINUTES • TOTAL: 50 MINUTES • SERVES: 6

Ingredients

FOR THE MARINADE
¼ cup hoisin sauce
2 teaspoons minced garlic
1 teaspoon sesame oil
1 tablespoon soy sauce
1 teaspoon ground ginger
¼ cup water
FOR THE STIR-FRY

1 pound beef top sirloin steak, cut into 1-inch strips
½ cup diced red onion
1 green bell pepper, cut into 1-inch strips
1 red bell pepper, cut into 1-inch strips
1 yellow bell pepper, cut into 1-inch strips
1 pound of broccoli florets
1 teaspoon stir-fry oil

Instructions:

1. In a small mixing bowl, make the marinade. Mix together the hoisin sauce, garlic, sesame oil, soy sauce, ginger, and water. Add the steak, then cover and let marinate in the refrigerator for about 30 minutes. In a large bowl, combine the vegetables and the stir-fry oil and toss until the vegetables are thoroughly coated with oil. Place the vegetables in the air fryer basket.
2. Set the temperature of your Innsky AF to 200°F. Set the timer and steam for 2 minutes. Check and make sure that the vegetables are soft. If not, add another 2 to 3 minutes. Once the vegetables are soft, transfer them to a large bowl and place the meat into the air fryer. Set the temperature of your Innsky AF to 360°F. Set the timer and fry for 4 minutes. Check the meat to make sure it's fully cooked. If not, add another 2 minutes. Pour the vegetables back into the Innsky air fryer basket and shake. Release the air fryer basket. Pour the meat and vegetables into a bowl and serve.

Per Serving: Calories: 284; Fat: 8g; Saturated fat: 1g; Carbohydrate: 24g; Fiber: 5g; Sugar: 9g; Protein: 31g; Iron: 2mg; Sodium: 525mg

Air Fryer Carne Asada Tacos

PREP: 5 MINUTES • COOK TIME: 14 MINUTES • TOTAL: 19 MINUTES • SERVES: 4

Ingredients

1½ pounds flank steak
Salt
Freshly ground black pepper
⅓ cup olive oil
⅓ cup freshly squeezed lime juice

½ cup chopped fresh cilantro
4 teaspoons minced garlic
1 teaspoon ground cumin
1 teaspoon chili powder

Instructions:

1. Spray the Innsky air fryer basket with olive oil. Place the flank steak in a large mixing bowl. Season with salt and pepper. Add the olive oil, lime juice, cilantro, garlic, cumin, and chili powder and toss to coat the steak. For the best flavor, let the steak marinate in the refrigerator for about 1 hour. Place the steak in the air fryer basket.
2. Set the temperature of your Innsky AF to 400°F. Set the timer and grill for 7 minutes. Using tongs, flip the steak. Reset the timer and grill for 7 minutes more. Cook the steak to your desired level of doneness. For medium-rare, cook to an internal temperature of 135°F, or for medium, 145°F. Be careful not to overcook the steak, or it will dry out. Let the steak rest for about 5 minutes, then cut into strips.

Per Serving: Calories: 371; Fat: 22g; Saturated fat: 1g; Carbohydrate: 4g; Fiber: 1g; Sugar: 1g; Protein: 38g; Iron: 1mg; Sodium: 67mg

Beefy 'n Cheesy Spanish Rice Casserole
PREP: 10 MINUTES • COOK TIME: 50 MINUTES • TOTAL: 60 MINUTES • SERVES: 3

Ingredients

2 tablespoons chopped green bell pepper
1 tablespoon chopped fresh cilantro
1/2-pound lean ground beef
1/2 cup water
1/2 teaspoon salt
1/2 teaspoon brown sugar
1/2 pinch ground black pepper

1/3 cup uncooked long grain rice
1/4 cup finely chopped onion
1/4 cup chile sauce
1/4 teaspoon ground cumin
1/4 teaspoon Worcestershire sauce
1/4 cup shredded Cheddar cheese
1/2 (14.5 ounce) can canned tomatoes

Instructions:

1   Lightly grease baking pan of air fryer with cooking spray. Add ground beef.

2   For 10 minutes, cook on 360°F. Halfway through cooking time, stir and crumble beef. Discard excess fat. Stir in pepper, Worcestershire sauce, cumin, brown sugar, salt, chile sauce, rice, water, tomatoes, green bell pepper, and onion. Mix well. Cover pan with foil and cook for 25 minutes. Stirring occasionally. Give it one last good stir, press down firmly and sprinkle cheese on top. Cook uncovered for 15 minutes at 390°F until tops are lightly browned. Serve and enjoy with chopped cilantro.

Air Fried Grilled Rib Eye with Herb Butter
PREP: 5 MINUTES PLUS 1 HOUR TO CHILL • COOK TIME: 8 MINUTES • TOTAL: 1 HOUR 13 MINUTES • SERVES: 4

Ingredients

FOR THE HERB BUTTER
1 cup (2 sticks) unsalted butter, at room temperature
1 garlic clove, roasted and peeled
1 tablespoon salt
1 teaspoon freshly ground black pepper
1 teaspoon minced shallot
1 teaspoon minced fresh parsley

1 teaspoon minced fresh sage
1 teaspoon minced fresh rosemary
FOR THE STEAK
4 (10- to 12-ounce) rib eye steaks
Salt
Freshly ground black pepper

Instructions:

1. TO MAKE THE HERB BUTTER: In a small mixing bowl, combine the butter, roasted garlic, salt, pepper, shallot, parsley, sage, and rosemary until the herbs are fully and evenly incorporated into the butter. Cover the butter and refrigerate for about 1 hour.
    TO MAKE THE STEAK: Preheat the Innsky air fryer to 400°F. Season the steaks with salt and pepper.

2. Place the steaks into the air fryer basket. Grill for 4 minutes.
    Using tongs, flip the steaks. Reset the timer and grill for 4 minutes more. Cook the steaks to your desired level of doneness. For medium-rare, cook to an internal temperature of 135°F, or for medium, 145°F. Place a dab of the butter on each steak.

    Per Serving: Calories: 586; Fat: 36g; Saturated fat: 19g; Carbohydrate: 5g; Fiber: 0g; Sugar: 0g; Protein: 61g; Iron: 8mg; Sodium: 821mg

Air Fryer Hamburgers
PREP: 5 MINUTES • COOK TIME: 8 MINUTES • TOTAL: 13 MINUTES • SERVES: 4

Ingredients

2 slices crustless bread
¼ cup milk
1½ pounds lean ground beef
1 teaspoon salt
½ teaspoon freshly ground black pepper
1 teaspoon minced garlic

4 hamburger buns
6 slices Foolproof Air Fryer Bacon, for topping (optional)
Lettuce, sliced tomato, and pickle, for topping (optional)

Instructions:

1   Cut the bread into 1-inch pieces. Put the bread pieces in a small mixing bowl and pour the milk over them. Let sit for about 5 minutes. In a medium mixing bowl, add the ground beef, the bread and milk mixture, salt, pepper, and minced garlic. Using your hands, mix well, making sure that the bread and milk mixture is broken down. Divide the meat into fourths and form into patties. You should have 4 (6-ounce) patties. Place the patties into the air fryer basket.

2   Set the temperature of your Innsky AF tor 400°F. Set the timer and grill for 4 minutes
Using a spatula, flip and cook for 4 minutes more. Assemble your hamburger by placing it in a hamburger bun and topping each burger with 1 or 2 slices of cooked bacon. Serve with a choice of toppings.

Per Serving: Calories: 558; Fat: 27g; Saturated fat: 9g; Carbohydrate: 29g; Fiber: 2g; Sugar: 4g; Protein: 50g; Iron: 6mg; Sodium: 1647mg

Beefy Steak Topped with Chimichurri Sauce
PREP: 5 MINUTES • COOK TIME: 60 MINUTES • TOTAL: 65 MINUTES • SERVES: 6

Ingredients

1 cup commercial chimichurri
3 pounds steak

Salt and pepper to taste

Instructions:

1   Place all ingredients in a Ziploc bag and marinate in the fridge for 2 hours.
Preheat the Innsky air fryer to 390°F. Place the grill pan accessory in the air fryer.

2   Grill the skirt steak for 20 minutes per batch. Flip the steak every 10 minutes for even grilling.

Bulgogi Beef
PREP: 10 MINUTES PLUS 30 MINUTES TO MARINATE • COOK TIME: 12 MINUTES • TOTAL: 52 MINUTES • SERVES: 6

Ingredients

1½ pounds sirloin or chuck steak, thinly sliced
3 green onions, each cut into 2 or 3 pieces
1 cup shredded carrots
3 tablespoons soy sauce
2 tablespoons brown sugar
2 tablespoons toasted sesame oil

2 tablespoons sesame seeds
2 cloves garlic, minced
½ teaspoon black pepper
Steamed rice, riced cauliflower, or green salad, for serving

Instructions:

1   In a resealable plastic bag, combine the sirloin, green onions, carrots, soy sauce, brown sugar, sesame oil, sesame seeds, garlic, and black pepper. Seal the bag and massage to coat all of the meat with the marinade. Marinate at room temperature for 30 minutes or in the refrigerator for up to 24 hours. Place the meat and vegetables in the air fryer basket, leaving behind as much of the marinade as possible; discard the marinade.

2   Set the Innsky air fryer to 400°F for 12 minutes, shaking halfway through the cooking time.
Serve with steamed rice or riced cauliflower, or over a green salad.

Air Fried Homemade Italian Meatballs
PREP: 10 MINUTES • COOK TIME: 10 MINUTES • TOTAL: 20 MINUTES • SERVES: 6

Ingredients

2 tablespoons olive oil, divided
1 onion, diced
1 pound ground beef
1 pound ground pork
⅓ cup plain bread crumbs
2 large eggs
¼ cup minced fresh parsley

2 teaspoons minced garlic
2 teaspoons salt
1 teaspoon freshly ground black pepper
½ teaspoon red pepper flakes
1 teaspoon Italian seasoning
3 tablespoons grated Parmesan cheese

Instructions:

1   Spray the Innsky air fryer basket with olive oil. In a small skillet, heat 1 tablespoon of olive oil over medium-low heat. Add the onions and sauté until soft. Let cool slightly. In a large bowl, mix together the cooked onions, ground beef, ground pork, bread crumbs, eggs, parsley, garlic, salt, black pepper, red pepper flakes, Italian seasoning, and grated Parmesan cheese. Form the mixture into meatballs about 1½ inches in diameter. Place the meatballs in the greased Innsky air fryer basket in a single layer.

2   Set the temperature of your Innsky AF to 400°F. Set the timer and fry for 5 minutes. Using tongs, flip the meatballs. Reset the timer and fry for 5 minutes more.

Beef & veggie Spring Rolls
PREP: 5 MINUTES • COOK TIME: 12 MINUTES • TOTAL: 55 MINUTES • SERVES: 10

Ingredients

2-ounce Asian rice noodles
1 tablespoon sesame oil
7-ounce ground beef
1 small onion, chopped
3 garlic cloves, crushed

1 cup fresh mixed vegetables
1 teaspoon soy sauce
1 packet spring roll skins
2 tablespoons water
Olive oil, as required

Instructions:

1   Soak the noodles in warm water till soft.
Drain and cut into small lengths. In a pan heat the oil and add the onion and garlic and sauté for about 4-5 minutes. Add beef and cook for about 4-5 minutes.
Add vegetables and cook for about 5-7 minutes or till cooked through.
Stir in soy sauce and remove from the heat.
Immediately, stir in the noodles and keep aside till all the juices have been absorbed.
Preheat the Innsky air fryer to 350 degrees F. and preheat the oven to 350 degrees F also.
Place the spring rolls skin onto a smooth surface.
Add a line of the filling diagonally across.
Fold the top point over the filling and then fold in both sides.
On the final point brush it with water before rolling to seal.
Brush the spring rolls with oil.

2   Arrange the rolls in batches in the Air fryer and Cook for about 8 minutes.
Repeat with remaining rolls.
Now, place spring rolls onto a baking sheet.
Bake for about 6 minutes per side.

Old-Fashioned Air Fryer Meatloaf
PREP: 10 MINUTES • COOK TIME: 15 MINUTES • TOTAL: 25 MINUTES • SERVES: 4

Ingredients

2 tablespoons unsalted butter
½ cup diced onion
½ cup diced green bell pepper
1 pound lean ground beef
1 pound ground pork
2 large eggs

1 tablespoon Worcestershire sauce
1 tablespoon soy sauce
1 teaspoon salt
½ teaspoon freshly ground black pepper
1 cup bread crumbs, divided
⅓ cup ketchup, divided

Instructions:

1   Spray 3 mini loaf pans with cooking spray. Heat the butter in a medium sauté pan or skillet over medium-low heat. Add the onions and green peppers and sauté until both are tender. Let the onions and green peppers cool to room temperature. Meanwhile, in a large bowl, combine the beef and pork. Add the eggs and mix well, then mix in the Worcestershire sauce, soy sauce, salt, and pepper. Add ½ cup of bread crumbs and mix well. Add more bread crumbs as needed. You only want to add enough bread crumbs to make the mixture stick together; if you add too much, your meatloaf will be dry. Mix in the green peppers and onions.
Evenly divide the meatloaf mixture among the 3 mini loaf pans. Gently pat the meat into the pan, so each pan is evenly filled. (Do not pack the pans, just make sure that the meat is evenly distributed in the pan.) Divide the ketchup among the loaf pans and spread it in an even layer over each loaf.

2   Set the temperature of your Innsky AF to 330°F. Set the timer and bake for 15 minutes. Check if the internal temperature of each meatloaf has reached 160°F. Add more time if needed. Using silicone oven mitts, remove the mini pans from the air fryer and let rest for about 5 minutes before serving.

Per Serving: Calories: 618; Fat: 34g; Saturated fat: 14g; Carbohydrate: 28g; Fiber: 2g; Sugar: 8g; Protein: 50g; Iron: 5mg; Sodium: 929mg

Beef Ribeye Steak
PREP: 5 MINUTES • COOK TIME: 20 MINUTES • TOTAL: 25 MINUTES • SERVES: 4

Ingredients

4 (8-ounce) ribeye steaks
1 tablespoon McCormick Grill Mates Montreal Steak Seasoning

Salt
Pepper

Instructions:

1   Season the steaks with the steak seasoning and salt and pepper to taste. Place 2 steaks in the Air fryer. You can use an accessory grill pan, a layer rack, or the standard Air fryer basket.

2   Cook for 4 minutes. Open the Air fryer and flip the steaks.
Cook for an additional 4 to 5 minutes. Check for doneness to determine how much additional cook time is need. Remove the cooked steaks from the Air fryer, then repeat steps 2 through 4 for the remaining 2 steaks. Cool before serving.

Per Serving: Calories: 293; Fat: 22g; Protein:23g; Fiber:0g

Baked Kibbeh

PREP: 15 MINUTES • COOK TIME: 28 MINUTES • TOTAL: 43 MINUTES • SERVES: 4

Ingredients

1 cup bulgur
2 medium onions, thinly sliced
½ cup pine nuts
2 tablespoons vegetable oil
2 teaspoons kosher salt

Vegetable oil spray
1pound ground lamb
4 teaspoons Lebanese Seven-Spice Mix
1 teaspoon ground cumin
Tzatziki, for serving

Instructions:

1 Rinse the bulgur well. Cover with cold water and let stand while you get the rest of the ingredients together. In a small bowl, combine the onions, pine nuts, oil, and 1 teaspoon of the salt. Spray a 7 × 3-inch round heatproof pan with vegetable oil spray. Place the onion mixture in the pan and place the pan in the air fryer basket.

2 Set the Innsky air fryer to 400°F for 8 minutes, stirring halfway through the cooking time. Meanwhile, in a stand mixer fitted with a paddle attachment, combine the drained bulgur, lamb, spice mix, cumin, and remaining 1 teaspoon salt. Beat until you have a smooth, sticky mixture. Remove the pan with the onion mixture from the air fryer and transfer the mixture to a small bowl. Spray the sides of the pan with oil spray. Spread half of the lamb mixture in the bottom of the pan. Top with all of the onion mixture. Top with the remaining lamb mixture, spreading to the sides of the pan.Place the pan in the air fryer basket. Set the Innsky air fryer to 350°F for 20 minutes, or until lamb is browned and crisp. Serve with tzatziki.

Chipotle Steak Tacos

PREP: 15 MINUTES PLUS 30 MINUTES TO MARINATE• COOK TIME: 8 MINUTES • TOTAL: 53 MINUTES • SERVES: 4

Ingredients

For the Beef Filling
1½ pounds flank steak, thinly sliced into long strips
2 tablespoons water
1 tablespoon olive oil
1 small red onion, diced
2 cloves garlic, crushed and peeled
1 canned chipotle chile in adobo sauce, plus 1 tablespoon adobo sauce from the can
1 tablespoon ancho chile powder

1 teaspoon ground cumin
1 teaspoon dried oregano
1½ teaspoons kosher salt
½ teaspoon black pepper
For Serving
8 lettuce leaves; or (6-inch) flour tortillas, warmed
½ cup crumbled cotija cheese
1 cup prepared salsa

Instructions:

1 *For the filling:* Place the beef strips in a resealable plastic bag. In a blender or food processor, combine the water, olive oil, onion, garlic, chipotle chile and adobo sauce, chile powder, cumin, oregano, salt, and pepper. Blend until smooth. Pour the marinade over the meat. Seal the bag and massage to coat. Marinate at room temperature for 30 minutes or in the refrigerator for up to 24 hours.

Using tongs, remove the beef strips from the bag (discard the marinade) and lay them flat in the air fryer basket, minimizing overlap as much as possible

Set the Innsky air fryer to 400°F for 8 minutes, turning the beef strips halfway through the cooking time.

To serve: Divide the meat among the lettuce or tortillas and top with cheese and salsa.

Air Fryer Roast Beef

PREP: 5 MINUTES • COOK TIME: 45 MINUTES • TOTAL: 50 MINUTES • SERVES: 6

Ingredients

Roast beef

1 tbsp. olive oil

Seasonings of choice

Instructions:

1 Ensure your air fryer is preheated to 160 degrees.
   Place roast in bowl and toss with olive oil and desired seasonings.
   Put seasoned roast into air fryer.
2 Set temperature to 160°F, and set time to 30 minutes and cook 30 minutes.
   Turn roast when the timer sounds and cook another 15 minutes.

Per Serving: Calories: 267; Fat: 8g; Protein:21g; Sugar:1g

German Rouladen–Style Steak

PREP: 20 MINUTES • COOK TIME: 15 MINUTES • TOTAL: 35 MINUTES • SERVES: 4

Ingredients

For the Onion Sauce

2 medium onions, cut into ½-inch-thick slices

Kosher salt and black pepper

½ cup sour cream

1 tablespoon tomato paste

2 teaspoons chopped fresh parsley

For the Rouladen

¼ cup Dijon mustard

1 pound flank or skirt steak, ¼ to ½ inch thick

1 teaspoon black pepper

4 slices bacon

¼ cup chopped fresh parsley

Instructions:

1 For the sauce: In a small bowl, mix together the onions with salt and pepper to taste. Place the onions in the air fryer basket. Set the Innsky air fryer to 400°F for 6 minutes, or until the onions are softened and golden brown. Set aside half of the onions to use in the rouladen. Place the rest in a small bowl and add the sour cream, tomato paste, parsley, ½ teaspoon salt, and ½ teaspoon pepper. Stir until well combined, adding 1 to 2 tablespoons of water, if necessary, to thin the sauce slightly. Set the sauce aside. For the rouladen: Evenly spread the mustard over the meat. Sprinkle with the pepper. Top with the bacon slices, reserved onions, and parsley. Starting at the long end, roll up the steak as tightly as possible, ending seam side down. Use 2 or 3 wooden toothpicks to hold the roll together. Using a sharp knife, cut the roll in half so that it better fits in the air fryer basket.
2 Place the steak, seam side down, in the air fryer basket. Set the Innsky air fryer to 400°F for 9 minutes. Use a meat thermometer to ensure the steak has reached an internal temperature of 145°F. Let the steak rest for 10 minutes before cutting into slices. Serve with the sauce.

Beef Korma

PREP: 10 MINUTES • COOK TIME: 20 MINUTES • TOTAL: 30 MINUTES • SERVES: 6

Ingredients

½ cup yogurt

1 tablespoon curry powder

1 tablespoon olive oil

1 onion, chopped

2 cloves garlic, minced

1 tomato, diced

½ cup frozen baby peas, thawed

Instructions:

1 In a medium bowl, combine the steak, yogurt, and curry powder. Stir and set aside. In a 6-inch metal bowl, combine the olive oil, onion, and garlic.
2 Cook for 3 to 4 minutes or until crisp and tender. Add the steak along with the yogurt and the diced tomato. Cook for 12 to 13 minutes or until steak is almost tender. Stir in the peas and cook for 2 to 3 minutes.

Italian Steak Rolls

PREP: 20 MINUTES PLUS 30 MINUTES TO MARINATE • COOK TIME: 9 MINUTES • TOTAL: 59 MINUTES • SERVES: 4

Ingredients

1 tablespoon vegetable oil
2 cloves garlic, minced
2 teaspoons dried Italian seasoning
1 teaspoon kosher salt
1 teaspoon black pepper

1 pound flank or skirt steak, ¼ to ½ inch thick
1 (10-ounce) package frozen spinach, thawed and squeezed dry
½ cup diced jarred roasted red pepper
1 cup shredded mozzarella cheese

Instructions:

1. In a large bowl, combine the oil, garlic, Italian seasoning, salt, and pepper. Whisk to combine. Add the steak to the bowl, turning to ensure the entire steak is covered with the seasonings. Cover and marinate at room temperature for 30 minutes or in the refrigerator for up to 24 hours. Lay the steak on a flat surface. Spread the spinach evenly over the steak, leaving a ¼-inch border at the edge. Evenly top each steak with the red pepper and cheese. Starting at a long end, roll up the steak as tightly as possible, ending seam side down. Use 2 or 3 wooden toothpicks to hold the roll together. Using a sharp knife, cut the roll in half so that it better fits in the air fryer basket.
   Place the steak roll, seam side down, in the air fryer basket.

2. Set the Innsky air fryer to 400°F for 9 minutes. Use a meat thermometer to ensure the steak has reached an internal temperature of 145°F. Let the steak rest for 10 minutes before cutting into slices to serve.

Cumin-Paprika Rubbed Beef Brisket

PREP: 5 MINUTES • COOK TIME: 2 HOURS • TOTAL: 2 HOURS, 5 MINUTES • SERVES: 12

Ingredients

¼ teaspoon cayenne pepper
1 ½ tablespoons paprika
1 teaspoon garlic powder
1 teaspoon ground cumin
1 teaspoon onion powder

2 teaspoons dry mustard
2 teaspoons ground black pepper
2 teaspoons salt
5 pounds brisket roast
5 tablespoons olive oil

Instructions:

1. Place all ingredients in a Ziploc bag and allow to marinate in the fridge for at least 2 hours. Preheat the Innsky air fryer for 5 minutes. Place the meat in a baking dish that will fit in the air fryer.

2. Place in the air fryer and cook for 2 hours at 350°F.

Per Serving: Calories: 269; Fat: 12.8g; Protein:35.6g; Fiber:2g

Chili-Espresso Marinated Steak

PREP: 5 MINUTES • COOK TIME: 50 MINUTES• TOTAL: 55 MINUTES • SERVES: 3

Ingredients

½ teaspoon garlic powder
1 ½ pounds beef flank steak
1 teaspoon instant espresso powder

2 tablespoons olive oil
2 teaspoons chili powder
Salt and pepper to taste

Instructions:

1. Preheat the Innsky air fryer to 390°F. Place the grill pan accessory in the air fryer. Make the dry rub by mixing the chili powder, salt, pepper, espresso powder, and garlic powder. Rub all over the steak and brush with oil.

2. Place on the grill pan and cook for 40 minutes. Halfway through the cooking time, flip the beef to cook evenly.

Per Serving: Calories: 249; Fat: 17g; Protein:20g; Fiber:2g

Kheema Burgers

PREP: 15 MINUTES • COOK TIME: 12 MINUTES • TOTAL: 27 MINUTES • SERVES: 4

Ingredients

For the Burgers
1 pound 85% lean ground beef or ground lamb
2 large eggs, lightly beaten
1 medium yellow onion, diced
¼ cup chopped fresh cilantro
1 tablespoon minced fresh ginger
3 cloves garlic, minced
2 teaspoons Garam Masala
1 teaspoon ground turmeric
½ teaspoon ground cinnamon

⅛ teaspoon ground cardamom
1 teaspoon kosher salt
1 teaspoon cayenne pepper
For the Raita Sauce
1 cup grated cucumber
½ cup sour cream
¼ teaspoon kosher salt
¼ teaspoon black pepper
For Serving
4 lettuce leaves, hamburger buns, or naan breads

Instructions:

1 *For the burgers:* In a large bowl, combine the ground beef, eggs, onion, cilantro, ginger, garlic, garam masala, turmeric, cinnamon, cardamom, salt, and cayenne. Gently mix until ingredients are thoroughly combined. Divide the meat into four portions and form into round patties. Make a slight depression in the middle of each patty with your thumb to prevent them from puffing up into a dome shape while cooking.

2 Place the patties in the air fryer basket. Set the Innsky air fryer to 350°F for 12 minutes. Use a meat thermometer to ensure the burgers have reached an internal temperature of 160°F (for medium).
Meanwhile, for the sauce: In a small bowl, combine the cucumber, sour cream, salt, and pepper.
To serve: Place the burgers on the lettuce, buns, or naan and top with the sauce.

Korean Beef Tacos

PREP: 10 MINUTES PLUS 30 MINUTES TO MARINATE • COOK TIME: 12 MINUTES • TOTAL: 52 MINUTES • SERVES: 6

Ingredients

2 tablespoons **gochujang** (Korean red chile paste)
2 cloves garlic, minced
2 teaspoons minced fresh ginger
2 tablespoons toasted sesame oil
1 tablespoon soy sauce
2 tablespoons sesame seeds
2 teaspoons sugar
½ teaspoon kosher salt

1½ pounds thinly sliced beef (chuck, rib eye, or sirloin)
1 medium red onion, sliced
12 (6-inch) flour tortillas, warmed; or lettuce leaves
½ cup chopped green onions
¼ cup chopped fresh cilantro (optional)
½ cup kimchi (optional)

Instructions:

1 In a small bowl, combine the **gochujang**, garlic, ginger, sesame oil, soy sauce, sesame seeds, sugar, and salt. Whisk until well combined. Place the beef and red onion in a resealable plastic bag and pour the marinade over. Seal the bag and massage to coat all of the meat and onion. Marinate at room temperature for 30 minutes or in the refrigerator for up to 24 hours.

2 Place the meat and onion in the air fryer basket, leaving behind as much of the marinade as possible; discard the marinade. Set the Innsky air fryer to 400°F for 12 minutes, shaking halfway through the cooking time.
To serve, place meat and onion in the tortillas. Top with the green onions and the cilantro and kimchi, if using, and serve.

Sugar-And-Spice Beef Empanadas
PREP: 15 MINUTES • COOK TIME: 15 MINUTES • TOTAL: 30 MINUTES • SERVES: 4

Ingredients
6 ounces of raw, lean ground beef
¼ cup of raw white onions, sliced and finely diced
1 teaspoon of cinnamon
½ teaspoon of nutmeg

½ teaspoon of ground cloves
1 small pinch of brown sugar
2 teaspoons of red chilli powder
pre-made empanada dough shells

Instructions:

1   In a deep stovetop saucepan, crumble and cook the ground beef at medium heat. Add in the onions, stirring continuously with a wooden spoon, then add the cinnamon, nutmeg and cloves. Break up the ground beef as it cooks, so it doesn't form large clumps. Remove the saucepan from the stovetop as soon as the beef is fully cooked, the onions are soft, and the spices are releasing their fragrances. Do not overcook, you want the meat to remain moist and juicy. Cover the saucepan and let stand on a heatsafe surface for a few minutes. Lay empanada shells flat on a clean counter. Spoon the spiced cooked beef from the saucepan into the empanada shells – a heaping spoonful on each, though not so much that the mixture spills over the edges. Fold the empanada shells over so that the spiced beef is fully covered. Seal edges with water and press down with a fork to secure. Sprinkle brown sugar over the still-wet seams of the empanadas, for an extra sweet crunch. Cover the basket of the Air fryer with a lining of tin foil, leaving the edges uncovered to allow air to circulate through the basket.

2   Place the empanadas in the foil-lined Innsky air fryer basket and set at 350 degrees for 15 minutes. Halfway through, slide the frying basket out and flip the empanadas using a spatula. Remove when golden, and serve directly from the basket onto plates.

Beef & Lemon Schnitzel for One
PREP: 5 MINUTES • COOK TIME: 12 MINUTES • TOTAL: 17 MINUTES • SERVES: 1

Ingredients
2 Tbsp Oil
2–3 oz Breadcrumbs
1 Whisked Egg in a Saucer/Soup Plate

1 Beef Schnitzel
1 Freshly Picked Lemon

Instructions:

1   Mix the oil and breadcrumbs together until loose and crumbly. Dip the meat into the egg, then into the crumbs. Make sure that it is evenly covered.

2   Gently place in the Air fryer basket, and cook at 350° F (preheat if needed) until done. The timing will depend on the thickness of the schnitzel, but for a relatively thin one, it should take roughly 12 min. Serve with a lemon half and a garden salad.

Crispy Mongolian Beef

PREP: 5 MINUTES • COOK TIME: 10 MINUTES • TOTAL: 15 MINUTES • SERVES: 6

Ingredients

Olive oil
½ C. almond flour
2 pounds beef tenderloin or beef chuck, sliced into strips
Sauce:
½ C. chopped green onion
1 tsp. red chili flakes
1 tsp. almond flour

½ C. brown sugar
1 tsp. hoisin sauce
½ C. water
½ C. rice vinegar
½ C. low-sodium soy sauce
1 tbsp. chopped garlic
1 tbsp. finely chopped ginger
2 tbsp. olive oil

Instructions:

1   Toss strips of beef in almond flour, ensuring they are coated well. Add to the Air fryer.
2   Set temperature to 300°F, and set time to 10 minutes, and cook 10 minutes at 300 degrees. Meanwhile, add all sauce ingredients to the pan and bring to a boil. Mix well. Add beef strips to the sauce and cook 2 minutes. Serve over cauliflower rice!

Montreal Steak Burgers

PREP: 10 MINUTES • COOK TIME: 10 MINUTES • TOTAL: 20 MINUTES • SERVES: 4

Ingredients

1 teaspoon mustard seeds
1 teaspoon cumin seeds
1 teaspoon coriander seeds
1 teaspoon dried minced garlic
1 teaspoon dried red pepper flakes
1 teaspoon kosher salt

2 teaspoons black peppercorns
1 pound 85% lean ground beef
2 tablespoons Worcestershire sauce
4 hamburger buns
Mayonnaise

Instructions:

1   In a mortar and pestle, combine the mustard seeds, cumin seeds, coriander seeds, dried garlic, pepper flakes, salt, and peppercorns. Roughly crush the seeds, stopping before you make a fine powder. In a large bowl, combine the spice mixture with the ground beef and Worcestershire sauce. Gently mix until well combined. Divide the meat into four portions and form into round patties. Make a slight depression in the middle of each patty with your thumb to prevent them from puffing up into a dome shape while cooking.
2   Place the patties in the air fryer basket. Set the Innsky air fryer to 350°F for 10 minutes. Use a meat thermometer to ensure the burgers have reached an internal temperature of 160°F (for medium).
To serve, place the burgers on the buns and top with mayonnaise.

Crispy Beef Schnitzel

PREP: 5 MINUTES • COOK TIME: 12 MINUTES • TOTAL: 17 MINUTES • SERVES: 1

Ingredients

1 beef schnitzel
Salt and ground black pepper, to taste
2 tablespoons olive oil

1/3 cup breadcrumbs
1 egg, whisked

Instructions:

Season the schnitzel with salt and black pepper.
In a mixing bowl, combine the oil and breadcrumbs. In another shallow bowl, beat the egg until frothy. Dip the schnitzel in the egg; then, dip it in the oil mixture.
Air-fry at 350 degrees F for 12 minutes. Enjoy.

Nigerian Peanut-Crusted Flank Steak

PREP: 15 MINUTES PLUS 30 MINUTES TO MARINATE • COOK TIME: 8 MINUTES • TOTAL: 53 MINUTES • SERVES: 4

Ingredients

For the Suya Spice Mix
¼ cup dry-roasted peanuts
1 teaspoon cumin seeds
1 teaspoon garlic powder
1 teaspoon smoked paprika
½ teaspoon ground ginger

1 teaspoon kosher salt
½ teaspoon cayenne pepper
For the Steak
1 pound flank steak
2 tablespoons vegetable oil

Instructions:

1 For the spice mix: In a clean coffee grinder or spice mill, combine the peanuts and cumin seeds. Process until you get a coarse powder. (Do not overprocess or you will wind up with peanut butter! Alternatively, you can grind the cumin with ⅓ cup ready-made peanut powder)

   Pour the peanut mixture into a small bowl, add the garlic powder, paprika, ginger, salt, and cayenne, and stir to combine. This recipe makes about ½ cup suya spice mix. Store leftovers in an airtight container in a cool, dry place for up to 1 month.

   For the steak: Cut the flank steak into ½-inch-thick slices, cutting against the grain and at a slight angle. Place the beef strips in a resealable plastic bag and add the oil and 2½ to 3 tablespoons of the spice mixture. Seal the bag and massage to coat all of the meat with the oil and spice mixture. Marinate at room temperature for 30 minutes or in the refrigerator for up to 24 hours.

2 Place the beef strips in the air fryer basket. Set the Innsky air fryer to 400°F for 8 minutes, turning the strips halfway through the cooking time.

   Transfer the meat to a serving platter. Sprinkle with additional spice mix, if desired.

Philly Cheesesteaks

PREP: 20 MINUTES • COOK TIME: 20 MINUTES • TOTAL: 40 MINUTES • SERVES: 2

Ingredients

12 ounces boneless rib-eye steak, sliced as thinly as possible
½ teaspoon Worcestershire sauce
½ teaspoon soy sauce
Kosher salt and black pepper
½ small onion, halved and thinly sliced

½ green bell pepper, stemmed, seeded, and thinly sliced
1 tablespoon vegetable oil
1 tablespoon butter, softened (optional)
2 soft hoagie rolls, split three-fourths of the way through
2 slices provolone cheese, halved

Instructions:

1 In a medium bowl, combine the meat, Worcestershire sauce, soy sauce, and salt and pepper to taste. Toss until the meat is evenly coated; set aside. In another medium bowl, combine the onion, bell pepper, and oil. Season to taste with salt and pepper. Toss until the vegetables are evenly coated.

2 Place the meat and vegetables in the air fryer basket. Set the Innsky air fryer to 400°F for 15 minutes, or until the meat and vegetables are cooked through, tossing once or twice. Transfer the meat and vegetables to a plate and cover lightly with foil; set aside.

   If using butter: Spread the insides of the rolls with the butter. Place the rolls in the air fryer basket, top sides down. Set the Innsky air fryer to 400°F for 3 minutes, or until the rolls are lightly toasted. Remove the rolls from the basket.

   Divide the meat and vegetables between the two rolls. Top each with cheese. Place in the air fryer basket. Set the Innsky air fryer to 400°F for 2 minutes, or until the cheese melts.

Simple Steak
PREP: 6MINUTES • COOK TIME: 14 MINUTES • TOTAL: 20 MINUTES • SERVES: 2

Ingredients

½ pound quality cuts steak

Salt and freshly ground black pepper, to taste

Instructions:

1. Preheat the Innsky air fryer to 390 degrees F.
   Rub the steak with salt and pepper evenly.
2. Place the steak in the Innsky air fryer basket and cook for about 14 minutes crispy.

Spicy Flank Steak with Zhoug
PREP: 20 MINUTES PLUS 1 HOUR TO MARINATE • COOK TIME: 8 MINUTES • TOTAL: 1 HOUR MINUTES • SERVES: 4

Ingredients

For the Marinade and Steak

½ cup dark beer or orange juice

¼ cup fresh lemon juice

3 cloves garlic, minced

2 tablespoons extra-virgin olive oil

2 tablespoons sriracha

2 tablespoons brown sugar

2 teaspoons ground cumin

2 teaspoons smoked paprika

1 tablespoon kosher salt

1 teaspoon black pepper

1½ pounds flank steak, trimmed and cut into 3 pieces

For the Zhoug

1 cup packed fresh cilantro leaves

2 cloves garlic, peeled

2 jalapeño or serrano chiles, stemmed and coarsely chopped

½ teaspoon ground cumin

¼ teaspoon ground coriander

¼ teaspoon kosher salt

2 to 4 tablespoons extra-virgin olive oil

Instructions:

1. *For the marinade and steak:* In a small bowl, whisk together the beer, lemon juice, garlic, olive oil, sriracha, brown sugar, cumin, paprika, salt, and pepper. Place the steak in a large resealable plastic bag. Pour the marinade over the steak, seal the bag, and massage the steak to coat. Marinate in the refrigerator for 1 hour or up to 24 hours, turning the bag occasionally. *Meanwhile, for the zhoug:* In a food processor, combine the cilantro, garlic, jalapeños, cumin, coriander, and salt. Process until finely chopped. Add 2 tablespoons olive oil and pulse to form a loose paste, adding up to 2 tablespoons more olive oil if needed. Transfer the zhoug to a glass container. Cover and store in the refrigerator until 30 minutes before serving if marinating more than 1 hour.

2. Remove the steak from the marinade and discard the marinade. Place the steak in the Innsky air fryer basket and Set the Innsky air fryer to 400°F for 8 minutes. Use a meat thermometer to ensure the steak has reached an internal temperature of 150°F (for medium). Transfer the steak to a cutting board and let rest for 5 minutes. Slice the steak across the grain and serve with the zhoug.

Garlic-Cumin And Orange Juice Marinated Steak
PREP: 6 MINUTES • COOK TIME: 60 MINUTES • TOTAL: 66 MINUTES • SERVES: 4

Ingredients

¼ cup orange juice

1 teaspoon ground cumin

2 pounds skirt steak, trimmed from excess fat

2 tablespoons lime juice

2 tablespoons olive oil

4 cloves of garlic, minced

Salt and pepper to taste

Instructions:

1. Place all ingredients in a mixing bowl and allow to marinate in the fridge for at least 2 hours. Preheat the Innsky air fryer to 390°F. Place the grill pan accessory in the air fryer.
2. Grill for 15 minutes per batch and flip the beef every 8 minutes for even grilling. Meanwhile, pour the marinade on a saucepan and allow to simmer for 10 minutes or until the sauce thickens. Slice the beef and pour over the sauce.

Beef with Beans
PREP: 10 MINUTES • COOK TIME: 13 MINUTES • TOTAL: 23 MINUTES • SERVES: 8

Ingredients

12 Ozs Lean Steak
1 Onion, sliced
1 Can Chopped Tomatoes
3/4 Cup Beef Stock

4 Tsp Fresh Thyme, chopped
1 Can Red Kidney Beans
Salt and Pepper to taste
Oven Safe Bowl

Instructions:

1. Preheat the Innsky air fryer to 390 degrees
   Trim the fat from the meat and cut into thin 1cm strips
   Add onion slices to the oven safe bowl and place in the Air fryer.
2. Cook for 3 minutes. Add the meat and continue cooking for 5 minutes.
   Add the tomatoes and their juice, beef stock, thyme and the beans and cook for an additional 5 minutes. Season with black pepper to taste.

Poblano Cheeseburgers with Avocado-Chipotle Mayo
PREP: 30 MINUTES • COOK TIME: 27 MINUTES • TOTAL: 57 MINUTES • SERVES: 4

Ingredients

For the Burgers
1 poblano pepper
Extra-virgin olive oil
1 pound 85% lean ground beef
2 cloves garlic, minced
Kosher salt and black pepper
4 slices Monterey Jack cheese
For the Avocado-Chipotle Mayonnaise
1 small ripe avocado, pitted, peeled, and cut into chunks

½ cup mayonnaise
Juice from ½ lime
2 canned chipotle peppers in adobo sauce, sliced in half lengthwise, seeded, and finely chopped
2 teaspoons chopped fresh cilantro
For Serving
4 lettuce leaves; or 4 hamburger buns, lightly toasted
1 red onion, sliced
Ripe tomatoes, sliced
Bibb lettuce

Instructions:

1. *For the burgers:* Brush the poblano pepper with a little bit of oil and place in the air fryer basket. Set the Innsky air fryer to 375°F for 15 minutes, turning the pepper halfway through the cooking time. When the pepper is soft, wrinkled, and charred, remove it from the air fryer and cover with a clean dish towel. Let stand for 5 minutes to steam. When the pepper is cool enough to handle, remove the skin, stem, and seeds and then dice the pepper.
   In a large bowl, gently mix the diced poblano, ground beef, and garlic until well combined. Shape into four patties and season both sides with salt and pepper to taste. Make a slight depression in the middle of each patty with your thumb to prevent them from puffing up into a dome shape while cooking.

2. Arrange the patties in the air fryer basket. Set the Innsky air fryer to 350°F for 12 minutes. Place a slice of cheese on each burger during the last minute or two of the cooking time. Use a meat thermometer to ensure the burgers have reached an internal temperature of 160°F (for medium).
   Meanwhile, for the mayonnaise: In a blender combine the avocado, mayonnaise, lime juice, chipotle peppers, and cilantro. Blend until smooth.
   To serve: Slather some mayo on a lettuce leaf or a bottom bun. Top with a patty, then add the onion, tomato, and lettuce (and finish with the top bun if using).

Beef Taco Fried Egg Rolls
PREP: 10 MINUTES • COOK TIME: 12 MINUTES • TOTAL: 25 MINUTES • SERVES: 8

Ingredients
1 tsp. cilantro
2 chopped garlic cloves
1 tbsp. olive oil
1 C. shredded Mexican cheese
½ packet taco seasoning

½ can cilantro lime rotel
½ chopped onion
16 egg roll wrappers
1pound lean ground beef

Instructions:
1   Ensure that your air fryer is preheated to 400 degrees.
Add onions and garlic to a skillet, cooking till fragrant. Then add taco seasoning, pepper, salt, and beef, cooking till beef is broke up into tiny pieces and cooked thoroughly.
Add rotel and stir well. Lay out egg wrappers and brush with water to soften a bit.
Load wrappers with beef filling and add cheese to each. Fold diagonally to close and use water to secure edges.
Brush filled egg wrappers with olive oil and add to the air fryer.
2   Set temperature to 400°F, and set time to 8 minutes. Cook 8 minutes, flip, and cook another 4 minutes. Served sprinkled with cilantro.

Taco Meatballs
PREP: 15 MINUTES • COOK TIME: 10 MINUTES • TOTAL: 25 MINUTES • SERVES: 4

Ingredients
For the Meatballs
1 pound 85% lean ground beef
½ cup shredded Mexican cheese blend
1 large egg
¼ cup finely minced onion
¼ cup chopped fresh cilantro
3 cloves garlic, minced

2½ tablespoons taco seasoning
1 teaspoon kosher salt
1 teaspoon black pepper
For the Sauce
¼ cup sour cream
½ cup salsa
1 to 2 teaspoons Cholula hot sauce or Sriracha

Instructions:
1   *For the meatballs:* In the bowl of a stand mixer fitted with the paddle attachment, combine the ground beef, cheese, egg, onion, cilantro, garlic, taco seasoning, salt, and pepper. Mix on low speed until all of the ingredients are incorporated, 2 to 3 minutes.
2   Form the mixture into 12 meatballs and arrange in a single layer in the air fryer basket. Set the Innsky air fryer to 400°F for 10 minutes. Use a meat thermometer to ensure the meatballs have reached an internal temperature of 160°F (for medium).
Meanwhile, for the sauce: In a small bowl, combine the sour cream, salsa, and hot sauce. Stir until well combined.
Transfer the meatballs to a serving bowl. Ladle the sauce over the meatballs and serve.

Wonton Meatballs

PREP: 15 MINUTES • COOK TIME: 10 MINUTES • TOTAL: 25 MINUTES • SERVES: 4

Ingredients

1 pound ground pork
2 large eggs
¼ cup chopped green onions (white and green parts)
¼ cup chopped fresh cilantro or parsley
1 tablespoon minced fresh ginger

3 cloves garlic, minced
2 teaspoons soy sauce
1 teaspoon oyster sauce
½ teaspoon kosher salt
1 teaspoon black pepper

Instructions:

1   In the bowl of a stand mixer fitted with the paddle attachment, combine the pork, eggs, green onions, cilantro, ginger, garlic, soy sauce, oyster sauce, salt, and pepper. Mix on low speed until all of the ingredients are incorporated, 2 to 3 minutes. Form the mixture into 12 meatballs and arrange in a single layer in the air fryer basket.

2   Set the Innsky air fryer to 350°F for 10 minutes. Use a meat thermometer to ensure the meatballs have reached an internal temperature of 145°F. Transfer the meatballs to a bowl and serve.

Swedish Meatballs

PREP: 10 MINUTES • COOK TIME: 14 MINUTES • TOTAL: 24 MINUTES • SERVES: 4

Ingredients

For the meatballs
1 pound 93% lean ground beef
1 (1-ounce) packet Lipton Onion Recipe Soup & Dip
Mix
⅓ cup bread crumbs
1 egg, beaten

Salt
Pepper
For the gravy
1 cup beef broth
⅓ cup heavy cream
3 tablespoons all-purpose flour

Instructions:

1   In a large bowl, combine the ground beef, onion soup mix, bread crumbs, egg, and salt and pepper to taste. Mix thoroughly. Using 2 tablespoons of the meat mixture, create each meatball by rolling the beef mixture around in your hands. This should yield about 10 meatballs.

2   Place the meatballs in the Air fryer. It is okay to stack them. Cook for 14 minutes. While the meatballs cook, prepare the gravy. Heat a saucepan over medium-high heat. Add the beef broth and heavy cream. Stir for 1 to 2 minutes. Add the flour and stir. Cover and allow the sauce to simmer for 3 to 4 minutes, or until thick. Drizzle the gravy over the meatballs and serve.

Reuben Egg Rolls

PREP: 5 MINUTES • COOK TIME: 20 MINUTES • TOTAL: 25 MINUTES • SERVES: 6

Ingredients

Swiss cheese
Can of sauerkraut

Sliced deli corned beef
Egg roll wrappers

Instructions:

1   Cut corned beef and Swiss cheese into thin slices. Drain sauerkraut and dry well. Take egg roll wrapper and moisten edges with water. Stack center with corned beef and cheese till you reach desired thickness. Top off with sauerkraut. Fold corner closest to you over the edge of filling. Bring up sides and glue with water. Add to the Innsky air fryer basket and spritz with olive oil.

2   Set temperature to 400°F, and set time to 4 minutes. Cook 4 minutes at 400 degrees, then flip and cook another 4 minutes.

Rice and Meatball Stuffed Bell Peppers
PREP: 13 MINUTES • COOK TIME: 15 MINUTES • TOTAL: 28 MINUTES • SERVES: 4

Ingredients

4 bell peppers
1 tablespoon olive oil
1 small onion, chopped
2 cloves garlic, minced

1 cup frozen cooked rice, thawed
16 to 20 small frozen precooked meatballs, thawed
½ cup tomato sauce
3 tablespoons Dijon mustard

Instructions:

1   To prepare the peppers, cut off about ½ inch of the tops. Carefully remove the membranes and seeds from inside the peppers. Set aside.
    In a 6-by-6-by-2-inch pan, combine the olive oil, onion, and garlic.
2   Bake in the Air fryer for 2 to 4 minutes or until crisp and tender. Remove the vegetable mixture from the pan and set aside in a medium bowl.
    Add the rice, meatballs, tomato sauce, and mustard to the vegetable mixture and stir to combine. Stuff the peppers with the meat-vegetable mixture.
    Place the peppers in the Innsky air fryer basket and bake for 9 to 13 minutes or until the filling is hot and the peppers are tender.

Per Serving: Calories: 487; Fat: 21g; Protein:26g; Fiber:6g

Stir-Fried Steak and Cabbage
PREP: 15 MINUTES • COOK TIME: 10 MINUTES • TOTAL: 35 MINUTES • *SERVES: 4*

Ingredients

½ pound sirloin steak, cut into strips
2 teaspoons cornstarch
1 tablespoon peanut oil
2 cups chopped red or green cabbage

1 yellow bell pepper, chopped
2 green onions, chopped
2 cloves garlic, sliced
½ cup commercial stir-fry sauce

Instructions:

1   Toss the steak with the cornstarch and set aside.
    In a 6-inch metal bowl, combine the peanut oil with the cabbage.
2   Place in the basket and cook for 3 to 4 minutes.
    Remove the bowl from the basket and add the steak, pepper, onions, and garlic. Return to the Air fryer and cook for 3 to 5 minutes or until the steak is cooked to desired doneness and vegetables are crisp and tender.
    Add the stir-fry sauce and cook for 2 to 4 minutes or until hot. Serve over rice.

Per Serving: Calories: 180; Fat: 7g; Protein:20g; Fiber:2g

Pub Style Corned Beef Egg Rolls

PREP: 15 MINUTES • COOK TIME: 10 MINUTES • TOTAL: 35 MINUTES • SERVES: 10

Ingredients

Olive oil
½ C. orange marmalade
5 slices of Swiss cheese
4 C. corned beef and cabbage
1 egg
10 egg roll wrappers
Brandy Mustard Sauce:
1/16th tsp. pepper
2 tbsp. whole grain mustard

1 tsp. dry mustard powder
1 C. heavy cream
½ C. chicken stock
¼ C. brandy
¾ C. dry white wine
¼ tsp. curry powder
½ tbsp. cilantro
1 minced shallot
2 tbsp. ghee

Instructions:

1   To make mustard sauce, add shallots and ghee to skillet, cooking until softened. Then add brandy and wine, heating to a low boil. Cook 5 minutes for liquids to reduce. Add stock and seasonings. Simmer 5 minutes. Turn down heat and add heavy cream. Cook on low till sauce reduces and it covers the back of a spoon. Place sauce in the fridge to chill. Crack the egg in a bowl and set to the side. Lay out an egg wrapper with the corner towards you. Brush the edges with egg wash. Place 1/3 cup of corned beef mixture into the center along with 2 tablespoons of marmalade and ½ a slice of Swiss cheese. Fold the bottom corner over filling. As you are folding the sides, make sure they are stick well to the first flap you made. Place filled rolls into prepared air fryer basket. Spritz rolls with olive oil.

2   Set temperature to 390°F, and set time to 10 minutes. Cook 10 minutes at 390 degrees, shaking halfway through cooking. Serve rolls with Brandy Mustard sauce.

Per Serving: Calories: 415; Fat: 13g; Protein:38g; Sugar:4g

Air-Fried Philly Cheesesteak

PREP: 5 MINUTES • COOK TIME: 16 MINUTES • TOTAL: 21 MINUTES • SERVES: 6

Ingredients

Large hoagie bun, sliced in half
6 ounces of sirloin or flank steak, sliced into bite-sized pieces

½ white onion, rinsed and sliced
½ red pepper, rinsed and sliced
slices of American cheese

Instructions:

1   Set the Innsky air fryer to 320 degrees for 10 minutes.
Arrange the steak pieces, onions and peppers on a piece of tin foil, flat and not overlapping, and set the tin foil on one side of the air-fryer basket. The foil should not take up more than half of the surface; the juices from the steak and the moisture from the vegetables will mingle while cooking. Lay the hoagie-bun halves, crusty-side up and soft-side down, on the other half of the air-fryer.

2   After 10 minutes, the air fryer will shut off; the hoagie buns should be starting to crisp and the steak and vegetables will have begun to cook. Carefully, flip the hoagie buns so they are now crusty-side down and soft-side up; cover both sides with one slice each of American cheese. With a long spoon, gently stir the steak, onions and peppers in the foil to ensure even coverage. Set the Innsky air fryer to 360 degrees for 6 minutes.
After 6 minutes, when the fryer shuts off, the cheese will be perfectly melted over the toasted bread, and the steak will be juicy on the inside and crispy on the outside.
Remove the cheesy hoagie halves first, using tongs, and set on a serving plate; then cover one side with the steak, and top with the onions and peppers. Close with the other cheesy hoagie-half, slice into two pieces, and enjoy.

Herbed Roast Beef
PREP: 5 MINUTES • COOK TIME: 20 MINUTES • TOTAL: 25 MINUTES • SERVES: 6

Ingredients
½ tsp. fresh rosemary
1 tsp. dried thyme
¼ tsp. pepper

1 tsp. salt
4-pound top round roast beef
tsp. olive oil

Instructions:
1   Ensure your air fryer is preheated to 360 degrees.
    Rub olive oil all over beef.
    Mix rosemary, thyme, pepper, and salt together and proceed to rub all sides of beef with spice mixture.
    Place seasoned beef into air fryer.
2   Set temperature to 360°F, and set time to 20 minutes.
    Allow roast to rest 10 minutes before slicing to serve.
Per Serving: Calories: 502; Fat: 18g; Protein:48g; Sugar:2g

Tender Beef with Sour Cream Sauce
PREP: 5 MINUTES • COOK TIME: 12 MINUTES • TOTAL: 17 MINUTES • SERVES: 2

Ingredients
9 ounces tender beef, chopped
1 cup scallions, chopped
2 cloves garlic, smashed
3/4 cup sour cream

3/4 teaspoon salt
1/4 teaspoon black pepper, or to taste
1/2 teaspoon dried dill weed

Instructions:

1   Add the beef, scallions, and garlic to the baking dish.
2   Cook for about 5 minutes at 390 degrees F.
    Once the meat is starting to tender, pour in the sour cream. Stir in the salt, black pepper, and dill.
    Now, cook 7 minutes longer.

Beef Empanadas
PREP: 5 MINUTES • COOK TIME: 20 MINUTES • TOTAL: 25 MINUTES •SERVES: 6

Ingredients
1 tsp. water
1 egg white

1 C. picadillo
8 Goya empanada discs (thawed)

Instructions:
1   Ensure your air fryer is preheated to 325. Spray basket with olive oil.
    Place 2 tablespoons of picadillo into the center of each disc. Fold disc in half and use a fork to seal edges. Repeat with all ingredients.
    Whisk egg white with water and brush tops of empanadas with egg wash.
    Add 2-3 empanadas to the Air fryer.
2   Set temperature to 325°F, and set time to 8 minutes, cook until golden. Repeat till you cook all filled empanadas.

Per Serving: Calories: 183; Fat: 5g; Protein:11g; Sugar:2g

Beef Pot Pie
PREP: 5 MINUTES • COOK TIME: 90 MINUTES • TOTAL: 95 MINUTES • SERVES: 2

Ingredients

1 tablespoon olive oil
1 pound beef stewing steak, cubed
1 large onion, chopped
1 tablespoon tomato puree
1 can ale

Warm water, as required
2 beef bouillon cubes
Salt and freshly ground black pepper, to taste
1 tablespoon plain flour plus more for dusting
1 prepared short crust pastry

Instructions:

1   In a pan, heat oil on medium heat. Add steak and cook for about 4-5 minutes. Add onion and cook for about 4-5 minutes. Add tomato puree and cook for about 2-3 minutes. In a jug, add the ale and enough water to double the mixture.
Add the ale mixture, cubes, salt and black pepper in the pan with beef and bring to a boil on high heat. Reduce the heat to low and simmer for about 1 hour. In a bowl, mix together flour and 3 tablespoons of warm water. Slowly, add the flour mixture in beef mixture, stirring continuously. Remove from heat and keep aside. Roll out the short crust pastry.
Line 2 ramekins with pastry and dust with flour. Divide the beef mixture in the ramekins evenly. Place extra pastry on top.

2   Preheat the Innsky air fryer to 390 degrees F, and Cook for about 10 minutes.
Now, Set the Innsky air fryer to 335 degrees F, and Cook for about 6 minutes more.

Bolognaise Sauce
PREP: 5 MINUTES • COOK TIME: 30 MINUTES • TOTAL: 35 MINUTES • SERVES: 2

Ingredients

13 Ozs Ground Beef
1 Carrot
1 Stalk of Celery
10 Ozs Diced Tomatoes

1/2 Onion
Salt and Pepper to taste
Oven safe bowl

Instructions:

1   Preheat the Innsky air fryer to 390 degrees. Finely dice the carrot, celery and onions. Place into the oven safe bowl along with the ground beef and combine well

2   Place the bowl into the Air fryer tray and cook for 12 minutes until browned.
Pour the diced tomatoes into the bowl and replace in the Air fryer. Season with salt and pepper, then cook for another 18 minutes. Serve over cooked pasta or freeze for later use.

Breaded Spam Steaks
PREP: 5 MINUTES • COOK TIME: 5 MINUTES • TOTAL: 10 MINUTES • SERVES: 2

Ingredients

12 Oz Can Luncheon Meat
1 Cup All Purpose Flour

2 Eggs, beaten
2 Cups Italian Seasoned Breadcrumbs

Instructions:

1   Preheat the Innsky air fryer to 380 degrees.
Cut the luncheon meat into 1/4 inch slices.
Gently press the luncheon meat slices into the flour to coat and shake off the excess flour. Dip into the beaten egg, then press into breadcrumbs.

2   Place the battered slices into the Air fryer tray and cook for 3 to 5 minutes until golden brown.
Serve with chili or tomato sauce

Air Fryer Burgers

Ingredients

1pound lean ground beef

1 tsp. dried parsley

½ tsp. dried oregano

½ tsp. pepper

½ tsp. salt

½ tsp. onion powder

½ tsp. garlic powder

Few drops of liquid smoke

1 tsp. Worcestershire sauce

Instructions:

1 Ensure your air fryer is preheated to 350 degrees.
Mix all seasonings together till combined. Place beef in a bowl and add seasonings. Mix well, but do not overmix. Make 4 patties from the mixture and using your thumb, making an indent in the center of each patty. Add patties to air fryer basket.

2 Set temperature to 350°F, and set time to 10 minutes, and cook 10 minutes.

Per serving: calories: 148; fat: 5g; protein:24g

Cheese-Stuffed Meatballs

Ingredients

⅓ cup soft bread crumbs

3 tablespoons milk

1 tablespoon ketchup

1 egg

½ teaspoon dried marjoram

Pinch salt

Freshly ground black pepper

1pound 95 percent lean ground beef

20 ½-inch cubes of cheese

Olive oil for misting

Instructions:

1 In a large bowl, combine the bread crumbs, milk, ketchup, egg, marjoram, salt, and pepper, and mix well. Add the ground beef and mix gently but thoroughly with your hands. Form the mixture into 20 meatballs. Shape each meatball around a cheese cube. Mist the meatballs with olive oil and put into the Air fryer basket.

2 Bake for 10 to 13 minutes or until the meatballs register 165°F on a meat thermometer.

Per Serving: Calories: 393; Fat: 17g; Protein:50g; Fiber:0g

Roasted Stuffed Peppers

Ingredients

4 ounces shredded cheddar cheese

½ tsp. pepper

½ tsp. salt

1 tsp. Worcestershire sauce

½ C. tomato sauce

8 ounces lean ground beef

1 tsp. olive oil

1 minced garlic clove

½ chopped onion

2 green peppers

Instructions:

1 Ensure your air fryer is preheated to 390 degrees. Spray with olive oil. Cut stems off bell peppers and remove seeds. Cook in boiling salted water for 3 minutes. Sauté garlic and onion together in a skillet until golden in color.
Take skillet off the heat. Mix pepper, salt, Worcestershire sauce, ¼ cup of tomato sauce, half of cheese and beef together. Divide meat mixture into pepper halves. Top filled peppers with remaining cheese and tomato sauce. Place filled peppers in the Air fryer.

2 Set temperature to 390°F, and set time to 20 minutes, bake 15-20 minutes.

Per Serving: Calories: 295; Fat: 8g; Protein:23g; Sugar:2g

Air Fried Steak Sandwich

PREP: 5 MINUTES • COOK TIME: 16 MINUTES • TOTAL: 21 MINUTES • SERVES: 4

Ingredients

Large hoagie bun, sliced in half

6 ounces of sirloin or flank steak, sliced into bite-sized pieces

½ tablespoon of mustard powder

½ tablespoon of soy sauce

1 tablespoon of fresh bleu cheese, crumbled

8 medium-sized cherry tomatoes, sliced in half

1 cup of fresh arugula, rinsed and patted dry

Instructions:

1   In a small mixing bowl, combine the soy sauce and onion powder; stir with a fork until thoroughly combined. Lay the raw steak strips in the soy-mustard mixture, and fully immerse each piece to marinate. Set the Innsky air fryer to 320 degrees for 10 minutes.

Arrange the soy-mustard marinated steak pieces on a piece of tin foil, flat and not overlapping, and set the tin foil on one side of the Air fryer basket. The foil should not take up more than half of the surface. Lay the hoagie-bun halves, crusty-side up and soft-side down, on the other half of the air-fryer.

2   After 10 minutes, the Air fryer will shut off; the hoagie buns should be starting to crisp and the steak will have begun to cook. Carefully, flip the hoagie buns so they are now crusty-side down and soft-side up; crumble a layer of the bleu cheese on each hoagie half. With a long spoon, gently stir the marinated steak in the foil to ensure even coverage. Set the Innsky air fryer to 360 degrees for 6 minutes.

After 6 minutes, when the fryer shuts off, the bleu cheese will be perfectly melted over the toasted bread, and the steak will be juicy on the inside and crispy on the outside.

Remove the cheesy hoagie halves first, using tongs, and set on a serving plate; then cover one side with the steak, and top with the cherry-tomato halves and the arugula. Close with the other cheesy hoagie-half, slice into two pieces, and enjoy.

Carrot and Beef Cocktail Balls

PREP: 5 MINUTES • COOK TIME: 20 MINUTES • TOTAL: 25 MINUTES • SERVES: 10

Ingredients

1 pound ground beef

2 carrots

1 red onion, peeled and chopped

2 cloves garlic

1/2 teaspoon dried rosemary, crushed

1/2 teaspoon dried basil

1 teaspoon dried oregano

1 egg

3/4 cup breadcrumbs

1/2 teaspoon salt

1/2 teaspoon black pepper, or to taste

1 cup plain flour

Instructions:

1   Place ground beef in a large bowl. In a food processor, pulse the carrot, onion and garlic; transfer the vegetable mixture to a large-sized bowl.

Then, add the rosemary, basil, oregano, egg, breadcrumbs, salt, and black pepper.

Shape the mixture into even balls; refrigerate for about 30 minutes. Roll the balls into the flour.

2   Then, air-fry the balls at 350 degrees F for about 20 minutes, turning occasionally; work with batches. Serve with toothpicks.

Beef Steaks with Beans

PREP: 5 MINUTES • COOK TIME: 10 MINUTES • TOTAL: 15 MINUTES • SERVES: 4

Ingredients

4 beef steaks, trim the fat and cut into strips
1 cup green onions, chopped
2 cloves garlic, minced
1 red bell pepper, seeded and thinly sliced
1 can tomatoes, crushed
1 can cannellini beans

3/4 cup beef broth
1/4 teaspoon dried basil
1/2 teaspoon cayenne pepper
1/2 teaspoon sea salt
1/4 teaspoon ground black pepper, or to taste

Instructions:

1   Add the steaks, green onions and garlic to the Air fryer basket.
2   Cook at 390 degrees F for 10 minutes, working in batches.
    Stir in the remaining ingredients and cook for an additional 5 minutes.

Air Fryer Beef Steak

PREP: 5 MINUTES • COOK TIME: 15 MINUTES • TOTAL: 20 MINUTES • SERVES: 4

Ingredients

1 tbsp. olive oil
Pepper and salt

2 pounds of ribeye steak

Instructions:

1   Season meat on both sides with pepper and salt.
    Rub all sides of meat with olive oil.
    Preheat air fryer to 356 degrees and spritz with olive oil.
2   Set temperature to 356°F, and set time to 7 minutes. Cook steak 7 minutes. Flip and cook an additional 6 minutes.
    Let meat sit 2-5 minutes to rest. Slice and serve with salad.

Per Serving: Calories: 233; Fat: 19g; Protein:16g; Sugar:0g

Mushroom Meatloaf

PREP: 5 MINUTES • COOK TIME: 25 MINUTES • TOTAL: 30 MINUTES • SERVES: 4

Ingredients

14-ounce lean ground beef
1 chorizo sausage, chopped finely
1 small onion, chopped
1 garlic clove, minced
2 tablespoons fresh cilantro, chopped

3 tablespoons breadcrumbs
1 egg
Salt and freshly ground black pepper, to taste
2 tablespoons fresh mushrooms, sliced thinly
3 tablespoons olive oil

Instructions:

1   Preheat the Innsky air fryer to 390 degrees F.
    In a large bowl, add all ingredients except mushrooms and mix till well combined.
    In a baking pan, place the beef mixture.
    With the back of spatula, smooth the surface.
    Top with mushroom slices and gently, press into the meatloaf.
    Drizzle with oil evenly.
2   Arrange the pan in the Innsky air fryer basket and cook for about 25 minutes.
    Cut the meatloaf in desires size wedges and serve.

Beef and Broccoli

PREP: 10 MINUTES • COOK TIME: 12 MINUTES • TOTAL: 25 MINUTES • SERVES: 4

Ingredients

1 minced garlic clove
1 sliced ginger root
1 tbsp. olive oil
1 tsp. almond flour
1 tsp. sweetener of choice
¾ pound round steak

1 tsp. low-sodium soy sauce
1/3 C. sherry
2 tsp. sesame oil
1/3 C. oyster sauce
1 pounds of broccoli

Instructions:

1   Remove stems from broccoli and slice into florets. Slice steak into thin strips.
    Combine sweetener, soy sauce, sherry, almond flour, sesame oil, and oyster sauce together, stirring till sweetener dissolves.
    Put strips of steak into the mixture and allow to marinate 45 minutes to 2 hours.
    Add broccoli and marinated steak to air fryer. Place garlic, ginger, and olive oil on top.
2   Set temperature to 400°F, and set time to 12 minutes. Cook 12 minutes at 400 degrees. Serve with cauliflower rice.

Per Serving: Calories: 384; Fat: 16g; Protein:19g; Sugar:4g

Air Fryer Beef Fajitas

PREP: 5 MINUTES • COOK TIME: 20 MINUTES • TOTAL: 25 MINUTES • SERVES: 6

Ingredients

Beef:
1/8 C. carne asada seasoning
2 pounds beef flap meat
Diet 7-Up
Fajita veggies:

1 tsp. chili powder
1-2 tsp. pepper
1-2 tsp. salt
2 bell peppers, your choice of color
1 onion

Instructions:

1   Slice flap meat into manageable pieces and place into a bowl. Season meat with carne seasoning and pour diet soda over meat. Cover and chill overnight. Ensure your air fryer is preheated to 380 degrees.
    Place a parchment liner into the Innsky air fryer basket and spray with olive oil. Place beef in layers into the basket. Cook 8-10 minutes, making sure to flip halfway through. Remove and set to the side. Slice up veggies and spray air fryer basket. Add veggies to the fryer and spray with olive oil.
2   Set temperature to 400°F, and set time to 10 minutes. Cook 10 minutes at 400 degrees, shaking 1-2 times during cooking process.
    Serve meat and veggies on wheat tortillas and top with favorite keto fillings.

Per Serving: Calories: 412; Fat: 21g; Protein:13g; Sugar:1Seafood Recipes

Coconut Shrimp

PREP: 5 MINUTES • COOK TIME: 10 MINUTES • TOTAL: 15 MINUTES • SERVES: 3

Ingredients

1 C. almond flour
1 C. panko breadcrumbs
1 tbsp. coconut flour

1 C. unsweetened, dried coconut
1 egg white
12 raw large shrimp

Instructions:

1   Put shrimp on paper towels to drain.
Mix coconut and panko breadcrumbs together. Then mix in coconut flour and almond flour in a different bowl. Set to the side. Dip shrimp into flour mixture, then into egg white, and then into coconut mixture. Place into air fryer basket. Repeat with remaining shrimp.
2   Set temperature to 350°F, and set time to 10 minutes. Turn halfway through cooking process.

Per Serving: Calories:213; Fat: 8g; Protein:15g; Sugar:3g

Air Fryer Cajun Shrimp

PREP: 5 MINUTES • COOK TIME: 6 MINUTES • TOTAL: 11 MINUTES • SERVES: 2

Ingredients

12 ounces uncooked medium shrimp, peeled and deveined
1 teaspoon cayenne pepper
1 teaspoon Old Bay seasoning

½ teaspoon smoked paprika
2 tablespoons olive oil
1 teaspoon salt

Instructions:

1   Preheat the Innsky air fryer to 390°F.
Meanwhile, in a medium mixing bowl, combine the shrimp, cayenne pepper, Old Bay, paprika, olive oil, and salt. Toss the shrimp in the oil and spices until the shrimp is thoroughly coated with both.
2   Place the shrimp in the air fryer basket. Set the timer and steam for 3 minutes. Remove the drawer and shake, so the shrimp redistribute in the basket for even cooking. Reset the timer and steam for another 3 minutes. Check that the shrimp are done. When they are cooked through, the flesh will be opaque. Add additional time if needed. Plate, serve, and enjoy!

Per Serving: Calories: 286; Fat: 16g; Saturated fat: 2g; Carbohydrate: 1g; Fiber: 0g; Sugar: 0g; Protein: 37g; Iron: 6mg; Sodium: 1868mg

Grilled Salmon

PREP: 5 MINUTES • COOK TIME: 10 MINUTES • TOTAL: 15 MINUTES • SERVES: 3

Ingredients

2 Salmon Fillets
1/2 Tsp Lemon Pepper
1/2 Tsp Garlic Powder
Salt and Pepper

1/3 Cup Soy Sauce
1/3 Cup Sugar
1 Tbsp Olive Oil

Instructions:

1   Season salmon fillets with lemon pepper, garlic powder and salt. In a shallow bowl, add a third cup of water and combine the olive oil, soy sauce and sugar. Place salmon the bowl and immerse in the sauce. Cover with cling film and allow to marinate in the refrigerator for at least an hour
2   Preheat the Innsky air fryer at 350 degrees.
Place salmon into the Air fryer and cook for 10 minutes or more until the fish is tender. Serve with lemon wedges

Homemade Air Fried Crab Cake Sliders
PREP: 5 MINUTES • COOK TIME: 10 MINUTES • TOTAL: 15 MINUTES • SERVES: 4

Ingredients
1 pound crabmeat, shredded
¼ cup bread crumbs
2 teaspoons dried parsley
1 teaspoon salt
½ teaspoon freshly ground black pepper
1 large egg

2 tablespoons mayonnaise
1 teaspoon dry mustard
4 slider buns
Sliced tomato, lettuce leaves, and rémoulade sauce, for topping

Instructions:
1. Spray the Innsky air fryer basket with olive oil or spray an air fryer–size baking sheet with olive oil or cooking spray.

   In a medium mixing bowl, combine the crabmeat, bread crumbs, parsley, salt, pepper, egg, mayonnaise, and dry mustard. Mix well.

   Form the crab mixture into 4 equal patties. (If the patties are too wet, add an additional 1 to 2 tablespoons of bread crumbs.)

   Place the crab cakes directly into the greased air fryer basket, or on the greased baking sheet set into the air fryer basket.

2. Set the temperature of your Innsky AF to 400°F. Set the timer and fry for 5 minutes.

   Flip the crab cakes. Reset the timer and fry the crab cakes for 5 minutes more.

   Serve on slider buns with sliced tomato, lettuce, and rémoulade sauce.

Per Serving: Calories: 294; Fat: 11g; Saturated fat: 2g; Carbohydrate: 20g; Fiber: 1g; Sugar: 3g; Protein: 27g; Iron: 2mg; Sodium: 1766mg

Air Fried Lobster Tails
PREP: 5 MINUTES • COOK TIME: 8 MINUTES • TOTAL: 13 MINUTES • SERVES: 2

Ingredients
2 tablespoons unsalted butter, melted
1 tablespoon minced garlic
1 teaspoon salt

1 tablespoon minced fresh chives
2 (4- to 6-ounce) frozen lobster tails

Instructions:
1. In a small mixing bowl, combine the butter, garlic, salt, and chives.

   Butterfly the lobster tail: Starting at the meaty end of the tail, use kitchen shears to cut down the center of the top shell. Stop when you reach the fanned, wide part of the tail. Carefully spread apart the meat and the shell along the cut line, but keep the meat attached where it connects to the wide part of the tail. Use your hand to gently disconnect the meat from the bottom of the shell. Lift the meat up and out of the shell (keeping it attached at the wide end). Close the shell under the meat, so the meat rests on top of the shell.

   Place the lobster in the Innsky air fryer basket and generously brush the butter mixture over the meat.

2. Set the temperature of your Innsky AF to 380°F. Set the timer and steam for 4 minutes.

   Open the air fryer and rotate the lobster tails. Brush them with more of the butter mixture. Reset the timer and steam for 4 minutes more. The lobster is done when the meat is opaque.

Per Serving: Calories: 255; Fat: 13g; Saturated fat: 7g; Carbohydrate: 2g; Fiber: 0g; Sugar: 0g; Protein: 32g; Iron: 0mg; Sodium: 1453mg

Crispy Paprika Fish Fillets
PREP: 5 MINUTES • COOK TIME: 15 MINUTES • TOTAL: 20 MINUTES • SERVES: 4

Ingredients

1/2 cup seasoned breadcrumbs
1 tablespoon balsamic vinegar
1/2 teaspoon seasoned salt
1 teaspoon paprika

1/2 teaspoon ground black pepper
1 teaspoon celery seed
2 fish fillets, halved
1 egg, beaten

Instructions:

1 Add the breadcrumbs, vinegar, salt, paprika, ground black pepper, and celery seeds to your food processor. Process for about 30 seconds.
   Coat the fish fillets with the beaten egg; then, coat them with the breadcrumbs mixture.
2 Cook at 350 degrees F for about 15 minutes.

Bacon Wrapped Shrimp
PREP: 5 MINUTES • COOK TIME: 5 MINUTES • TOTAL: 10 MINUTES • SERVES: 4

Ingredients

1¼ pound tiger shrimp, peeled and deveined
1 pound bacon

Instructions:

1 Wrap each shrimp with a slice of bacon.
   Refrigerate for about 20 minutes. Preheat the Innsky air fryer to 390 degrees F.
2 Arrange the shrimp in the Air fryer basket. Cook for about 5-7 minutes.

Bacon-Wrapped Scallops
PREP: 5 MINUTES • COOK TIME: 10 MINUTES • TOTAL: 15 MINUTES • SERVES: 4

Ingredients

16 sea scallops
8 slices bacon, cut in half
8 toothpicks

Salt
Freshly ground black pepper

Instructions:

1 Using a paper towel, pat dry the scallops.
   Wrap each scallop with a half slice of bacon. Secure the bacon with a toothpick.
   Place the scallops into the air fryer in a single layer. (You may need to cook your scallops in more than one batch.)
   Spray the scallops with olive oil, and season them with salt and pepper.
2 Set the temperature of your Innsky AF to 370°F. Set the timer and fry for 5 minutes.
   Flip the scallops.
   Reset your timer and cook the scallops for 5 minutes more.
   Using tongs, remove the scallops from the air fryer basket. Plate, serve, and enjoy!

Per Serving: Calories: 311; Fat: 17g; Saturated fat: 5g; Carbohydrate: 3g; Fiber: 0g; Sugar: 0g; Protein: 34g; Iron: 1mg; Sodium: 1110mg

Air Fryer Salmon
PREP: 5 MINUTES • COOK TIME: 10 MINUTES • TOTAL: 15 MINUTES • SERVES: 2

Ingredients

½ tsp. salt
½ tsp. garlic powder

½ tsp. smoked paprika
Salmon

Instructions:

1. Mix spices together and sprinkle onto salmon. Place seasoned salmon into the Air fryer.
2. Close crisping lid. Set temperature to 400°F, and set time to 10 minutes.

Per Serving: Calories: 185; Fat: 11g; Protein:21g; Sugar:0g

Lemon Pepper, Butter, And Cajun Cod
PREP: 5 MINUTES • COOK TIME: 12 MINUTES • TOTAL: 17 MINUTES • SERVES: 2

Ingredients

2 (8-ounce) cod fillets, cut to fit into the air fryer basket
1 tablespoon Cajun seasoning
½ teaspoon lemon pepper

1 teaspoon salt
½ teaspoon freshly ground black pepper
2 tablespoons unsalted butter, melted
1 lemon, cut into 4 wedges

Instructions:

1. Spray the Innsky air fryer basket with olive oil. Place the fillets on a plate. In a small mixing bowl, combine the Cajun seasoning, lemon pepper, salt, and pepper.
   Rub the seasoning mix onto the fish.
   Place the cod into the greased air fryer basket. Brush the top of each fillet with melted butter.
2. Set the temperature of your Innsky AF to 360°F. Set the timer and bake for 6 minutes. After 6 minutes, open up your air fryer drawer and flip the fish. Brush the top of each fillet with more melted butter.
   Reset the timer and bake for 6 minutes more. Squeeze fresh lemon juice over the fillets.

Per Serving: Calories: 283; Fat: 14g; Saturated fat: 7g; Carbohydrate: 0g; Fiber: 0g; Sugar: 0g; Protein: 40g; Iron: 0mg; Sodium: 1460mg

Steamed Salmon & Sauce
PREP: 5 MINUTES • COOK TIME: 10 MINUTES • TOTAL: 15 MINUTES • SERVES: 2

Ingredients

1 cup Water
2 x 6 oz Fresh Salmon
2 Tsp Vegetable Oil
A Pinch of Salt for Each Fish
½ cup Plain Greek Yogurt

½ cup Sour Cream
2 Tbsp Finely Chopped Dill (Keep a bit for garnishing)
A Pinch of Salt to Taste

Instructions:

1. Pour the water into the bottom of the fryer and start heating to 285° F.
   Drizzle oil over the fish and spread it. Salt the fish to taste.
2. Now pop it into the fryer for 10 min.
   In the meantime, mix the yogurt, cream, dill and a bit of salt to make the sauce. When the fish is done, serve with the sauce and garnish with sprigs of dill.

Salmon Patties

PREP: 5 MINUTES • COOK TIME: 10 MINUTES • TOTAL: 15 MINUTES • SERVES: 4

Ingredients

1 (14.75-ounce) can wild salmon, drained
1 large egg
¼ cup diced onion
½ cup bread crumbs

1 teaspoon dried dill
½ teaspoon freshly ground black pepper
1 teaspoon salt
1 teaspoon Old Bay seasoning

Instructions:

1   Spray the Innsky air fryer basket with olive oil. Put the salmon in a medium bowl and remove any bones or skin. Add the egg, onion, bread crumbs, dill, pepper, salt, and Old Bay seasoning and mix well. Form the salmon mixture into 4 equal patties. Place the patties in the greased air fryer basket.

2   Set the temperature of your Innsky AF to 370°F. Set the timer and grill for 5 minutes. Flip the patties. Reset the timer and grill the patties for 5 minutes more. Plate, serve, and enjoy.

Per Serving: Calories: 239; Fat: 9g; Saturated fat: 2g; Carbohydrate: 11g; Fiber: 1g; Sugar: 1g; Protein: 27g; Iron: 2mg; Sodium: 901mg

Sweet And Savory Breaded Shrimp

PREP: 5 MINUTES • COOK TIME: 20 MINUTES • TOTAL: 25 MINUTES • SERVES: 2

Ingredients

½ pound of fresh shrimp, peeled from their shells and rinsed
2 raw eggs
½ cup of breadcrumbs (we like Panko, but any brand or home recipe will do)
½ white onion, peeled and rinsed and finely chopped
1 teaspoon of ginger-garlic paste

½ teaspoon of turmeric powder
½ teaspoon of red chili powder
½ teaspoon of cumin powder
½ teaspoon of black pepper powder
½ teaspoon of dry mango powder
Pinch of salt

Instructions:

1   Cover the basket of the Air fryer with a lining of tin foil, leaving the edges uncovered to allow air to circulate through the basket.
Preheat the Innsky air fryer to 350 degrees. In a large mixing bowl, beat the eggs until fluffy and until the yolks and whites are fully combined. Dunk all the shrimp in the egg mixture, fully submerging. In a separate mixing bowl, combine the bread crumbs with all the dry ingredients until evenly blended. One by one, coat the egg-covered shrimp in the mixed dry ingredients so that fully covered, and place on the foil-lined air-fryer basket.

2   Set the air-fryer timer to 20 minutes. Halfway through the cooking time, shake the handle of the air-fryer so that the breaded shrimp jostles inside and fry-coverage is even. After 20 minutes, when the fryer shuts off, the shrimp will be perfectly cooked and their breaded crust golden-brown and delicious! Using tongs, remove from the air fryer and set on a serving dish to cool.

Healthy Fish and Chips
PREP: 5 MINUTES • COOK TIME: 15 MINUTES • TOTAL: 20 MINUTES • SERVES: 3

Ingredients
Old Bay seasoning
½ C. panko breadcrumbs
1 egg

2 tbsp. almond flour
4-6 ounce tilapia fillets
Frozen crinkle cut fries

Instructions:
1   Add almond flour to one bowl, beat egg in another bowl, and add panko breadcrumbs to the third bowl, mixed with Old Bay seasoning. Dredge tilapia in flour, then egg, and then breadcrumbs. Place coated fish in Air fryer along with fries.
2   Set temperature to 390°F, and set time to 15 minutes.

Per Serving: Calories: 219; Fat: 5g; Protein:25g; Sugar:1g

Indian Fish Fingers
PREP: 35 MINUTES • COOK TIME: 15 MINUTES • TOTAL: 50 MINUTES • SERVES: 4

Ingredients
1/2pound fish fillet
1 tablespoon finely chopped fresh mint leaves or any fresh herbs
1/3 cup bread crumbs
1 teaspoon ginger garlic paste or ginger and garlic powders
1 hot green chili finely chopped

1/2 teaspoon paprika
Generous pinch of black pepper
Salt to taste
3/4 tablespoons lemon juice
3/4 teaspoons garam masala powder
1/3 teaspoon rosemary
1 egg

Instructions:
1   Start by removing any skin on the fish, washing, and patting dry. Cut the fish into fingers. In a medium bowl mix together all ingredients except for fish, mint, and bread crumbs. Bury the fingers in the mixture and refrigerate for 30 minutes. Remove from the bowl from the fridge and mix in mint leaves. In a separate bowl beat the egg, pour bread crumbs into a third bowl. Dip the fingers in the egg bowl then toss them in the bread crumbs bowl.
2   Cook at 360 degrees for 15 minutes, toss the fingers halfway through.

Per Serving: Calories: 187; Fat: 7g; Protein:11g; Fiber:1g

Lemon, Garlic, And Herb Salmon
PREP: 5 MINUTES • COOK TIME: 10 MINUTES • TOTAL: 15 MINUTES • SERVES: 4

Ingredients
3 tablespoons unsalted butter
1 garlic clove, minced, or ½ teaspoon garlic powder
1 teaspoon salt
2 tablespoons freshly squeezed lemon juice
1 tablespoon minced fresh parsley

1 teaspoon minced fresh dill
1 teaspoon salt
½ teaspoon freshly ground black pepper
4 (4-ounce) salmon fillets

Instructions:
1   Line the Innsky air fryer basket with parchment paper.
In a small microwave-safe mixing bowl, combine the butter, garlic, salt, lemon juice, parsley, dill, salt, and pepper. Place the bowl in the microwave and cook on low until the butter is completely melted, about 45 seconds. Meanwhile, place the salmon fillets in the parchment-lined air fryer basket. Spoon the sauce over the salmon.
2   Set the temperature of your Innsky AF to 400°F. Set the timer and bake for 10 minutes. Since you don't want to overcook the salmon, begin checking for doneness at about 8 minutes. Salmon is done when the flesh is opaque and flakes easily when tested with a fork.

Shrimp Scampi

PREP: 5 MINUTES • COOK TIME: 8 MINUTES • TOTAL: 13 MINUTES • SERVES: 4

Ingredients

¼ cup unsalted butter (or butter-flavored coconut oil for dairy-free)

2 tablespoons fish stock or chicken broth

1 tablespoon lemon juice

2 cloves garlic, minced

2 tablespoons chopped fresh basil leaves

1 tablespoon chopped fresh parsley, plus more for garnish

1 teaspoon red pepper flakes

1pound large shrimp, peeled and deveined, tails removed

Fresh basil sprigs, for garnish

Instructions:

1   Preheat the Innsky air fryer to 350° F. Place the butter, fish stock, lemon juice, garlic, basil, parsley, and red pepper flakes in a 6 by 3-inch pan, stir to combine, and place in the air fryer.

2   Cook for 3 minutes, or until fragrant and the garlic has softened.
Add the shrimp and stir to coat the shrimp in the sauce. Cook for 5 minutes, or until the shrimp are pink, stirring after 3 minutes. Garnish with fresh basil sprigs and chopped parsley before serving. Store leftovers in an airtight container in the refrigerator for up to 4 days. Reheat in a preheated 400°F air fryer for about 3 minutes, until heated through.

Per serving: Calories 175; Fat 11g; Protein 18g; Total carbs 1g; Fiber 0.2g

Simple Scallops

PREP: 5 MINUTES • COOK TIME: 4 MINUTES • TOTAL: 9 MINUTES • SERVES: 2

Ingredients

12 medium sea scallops

1 teaspoon fine sea salt

¾ teaspoon ground black pepper, plus more for garnish if desired

Fresh thyme leaves, for garnish (optional)

Instructions:

1   Spray the Innsky air fryer basket with avocado oil. Preheat the Innsky air fryer to 390°F. Rinse the scallops and pat completely dry. Spray avocado oil on the scallops and season them with the salt and pepper.

2   Place them in the air fryer basket, spacing them apart (if you're using a smaller air fryer, work in batches if necessary). Cook for 2 minutes, then flip the scallops and cook for another 2 minutes, or until cooked through and no longer translucent. Garnish with ground black pepper and thyme leaves, if desired. Best served fresh. Store leftovers in an airtight container in the fridge for up to 3 days. Reheat in a preheated 350°F air fryer for 2 minutes, or until heated through.

Quick Paella

PREP: 7 MINUTES • COOK TIME: 15 MINUTES • TOTAL: 22 MINUTES • SERVES: 4

Ingredients

1 (10-ounce) package frozen cooked rice, thawed

1 (6-ounce) jar artichoke hearts, drained and chopped

¼ cup vegetable broth

½ teaspoon turmeric

½ teaspoon dried thyme

1 cup frozen cooked small shrimp

½ cup frozen baby peas

1 tomato, diced

Instructions:

1   In a 6-by-6-by-2-inch pan, combine the rice, artichoke hearts, vegetable broth, turmeric, and thyme, and stir gently.

2   Place in the Air fryer and bake for 8 to 9 minutes or until the rice is hot. Remove from the air fryer and gently stir in the shrimp, peas, and tomato. Cook for 5 to 8 minutes or until the shrimp and peas are hot and the paella is bubbling.

Tuna Melt Croquettes

PREP: 5 MINUTES • COOK TIME: 10 MINUTES • TOTAL: 15 MINUTES • SERVES: 1 Dozen Croquettes

Ingredients

2 (5-ounce) cans tuna, drained

1 (8-ounce) package cream cheese, softened

½ cup finely shredded cheddar cheese

2 tablespoons diced onions

2 teaspoons prepared yellow mustard

1 large egg

1½ cups pork dust or powdered Parmesan cheese

Fresh dill, for garnish (optional)

FOR SERVING (OPTIONAL):

Cherry tomatoes

Mayonnaise

Prepared yellow mustard

Instructions:

1   Preheat the Innsky air fryer to 400°F.
Make the patties: In a large bowl, stir together the tuna, cream cheese, cheddar cheese, onions, mustard, and egg until well combined. Place the pork dust in a shallow bowl.
Form the tuna mixture into twelve 1½-inch balls. Roll the balls in the pork dust and use your hands to press it into a thick crust around each ball. Flatten the balls into ½-inch-thick patties.

2   Working in batches to avoid overcrowding, place the patties in the air fryer basket, leaving space between them. Cook for 8 minutes, or until golden brown and crispy, flipping halfway through. Garnish the croquettes with fresh dill, if desired, and serve with cherry tomatoes and dollops of mayo and mustard on the side.
Store leftovers in an airtight container in the refrigerator for up to 4 days. Reheat in a preheated 400°F air fryer for about 3 minutes, until heated through.

Per serving: Calories 528; Fat 36g; Protein 48g; Total carbs 2g; Fiber 0.3g

Coconut Shrimp

PREP: 15 MINUTES • COOK TIME: 5 MINUTES • TOTAL: 20 MINUTES • SERVES: 4

Ingredients

1 (8-ounce) can crushed pineapple

½ cup sour cream

¼ cup pineapple preserves

2 egg whites

⅔ cup cornstarch

⅔ cup sweetened coconut

1 cup panko bread crumbs

1pound uncooked large shrimp, thawed if frozen, deveined and shelled

Olive oil for misting

Instructions:

1   Drain the crushed pineapple well, reserving the juice. In a small bowl, combine the pineapple, sour cream, and preserves, and mix well. Set aside. In a shallow bowl, beat the egg whites with 2 tablespoons of the reserved pineapple liquid. Place the cornstarch on a plate. Combine the coconut and bread crumbs on another plate. Dip the shrimp into the cornstarch, shake it off, then dip into the egg white mixture and finally into the coconut mixture. Place the shrimp in the Innsky air fryer basket and mist with oil.

2   Air-fry for 5 to 7 minutes or until the shrimp are crisp and golden brown.

Per Serving: Calories: 524; Fat: 14g; Protein:33g; Fiber:4g

3-Ingredient Air Fryer Catfish
PREP: 5 MINUTES • COOK TIME: 13 MINUTES • TOTAL: 20 MINUTES • SERVES: 4

Ingredients

1 tbsp. chopped parsley
1 tbsp. olive oil

¼ C. seasoned fish fry
4 catfish fillets

Instructions:

1 Ensure your air fryer is preheated to 400 degrees.
   Rinse off catfish fillets and pat dry. Add fish fry seasoning to Ziploc baggie, then catfish. Shake bag and ensure fish gets well coated. Spray each fillet with olive oil. Add fillets to air fryer basket.
2 Set temperature to 400°F, and set time to 10 minutes. Cook 10 minutes. Then flip and cook another 2-3 minutes.

Per Serving: Calories: 208; Fat: 5g; Protein:17g; Sugar:0.5g

Coconut Shrimp with Spicy Mayo
PREP: 10 MINUTES • COOK TIME: 6 MINUTES • TOTAL: 16 MINUTES • SERVES: 4

Ingredients

1pound large shrimp (about 2 dozen), peeled and deveined, tails on
Fine sea salt and ground black pepper
2 large eggs
1 tablespoon water
½ cup unsweetened coconut flakes
½ cup pork dust
SPICY MAYO:

½ cup mayonnaise
2 tablespoons beef or chicken broth
½ teaspoon hot sauce
½ teaspoon cayenne pepper
FOR SERVING (OPTIONAL):
Microgreens
Thinly sliced radishes

Instructions:

1 Spray the Innsky air fryer basket with avocado oil. Preheat the Innsky air fryer to 400°F.
   Season the shrimp well on all sides with salt and pepper.
   Crack the eggs into a shallow baking dish, add the water and a pinch each of salt and pepper, and whisk to combine. In another shallow baking dish, stir together the coconut flakes and pork dust until well combined. Dip one shrimp in the eggs and let any excess egg drip off, then dredge both sides of the shrimp in the coconut mixture. Spray the shrimp with avocado oil and place it in the air fryer basket. Repeat with the remaining shrimp, leaving space between them in the air fryer basket.
2 Cook the shrimp in the air fryer for 6 minutes, or until cooked through and no longer translucent, flipping halfway through.
   While the shrimp cook, make the spicy mayo: In a medium-sized bowl, stir together all the spicy mayo ingredients until well combined.
   Serve the shrimp on a bed of microgreens and thinly sliced radishes, if desired. Serve the spicy mayo on the side for dipping.
   Store leftovers in an airtight container in the refrigerator for up to 4 days. Reheat in a preheated 400°F air fryer for about 3 minutes, until heated through.

Per serving: Calories 360; Fat 28g; Protein 25g; Total carbs 2g; Fiber 1g

Crispy Crab Rangoon Patties with Sweet and Sour Sauce
PREP: 10 MINUTES • COOK TIME: 12 MINUTES • TOTAL: 22 MINUTES • SERVES: 8

Ingredients

1 pound canned lump crabmeat, drained
1 (8-ounce) package cream cheese, softened
1 tablespoon chopped fresh chives
1 large egg
1 teaspoon grated fresh ginger
1 clove garlic, smashed to a paste or minced
COATING:
1½ cups pork dust
DIPPING SAUCE:
½ cup chicken broth

⅓ cup coconut aminos or wheat-free tamari
⅓ cup Swerve confectioners'-style sweetener or equivalent amount of liquid or powdered sweetener
¼ cup tomato sauce
1 tablespoon coconut vinegar or apple cider vinegar
¼ teaspoon grated fresh ginger
1 clove garlic, smashed to a paste
Sliced green onions, for garnish (optional)
Fried Cauliflower Rice, for serving (optional)

Instructions:

1. Preheat the Innsky air fryer to 400°F.
   In a medium-sized bowl, gently mix all the ingredients for the patties, without breaking up the crabmeat. Form the crab mixture into 8 patties that are 2½ inches in diameter and ¾ inch thick.
   Place the pork dust in a shallow dish. Place each patty in the pork dust and use your hands to press the pork dust into the patties to form a crust.
2. Place the patties in a single layer in the air fryer, leaving space between them. (If you're using a smaller air fryer, work in batches if necessary.) Cook for 12 minutes, or until the crust is golden and crispy. While the patties cook, make the dipping sauce: In a large saucepan, whisk together all the sauce ingredients. Bring to a simmer over medium-high heat, then turn the heat down to medium until the sauce has reduced and thickened, about 5 minutes. Taste and adjust the seasonings as desired. Place the patties on a serving platter, drizzle with the dipping sauce, and garnish with sliced green onions, if desired. Serve the remaining dipping sauce on the side. Serve with fried cauliflower rice, if desired. Store leftovers in an airtight container in the refrigerator for up to 3 days. Reheat the patties in a preheated 400°F air fryer for 4 minutes, or until crispy on the outside and heated through.

Per serving: Calories 411; Fat 30g; Protein 35g; Total carbs 4g; Fiber 3g

Tuna Veggie Stir-Fry
PREP: 5 MINUTES • COOK TIME: 12 MINUTES • TOTAL: 17 MINUTES • SERVES: 4

Ingredients

1 tablespoon olive oil
1 red bell pepper, chopped
1 cup green beans, cut into 2-inch pieces
1 onion, sliced
2 cloves garlic, sliced

2 tablespoons low-sodium soy sauce
1 tablespoon honey
½ pound fresh tuna, cubed

Instructions:

1 In a 6-inch metal bowl, combine the olive oil, pepper, green beans, onion, and garlic.
2 Cook in the Air fryer for 4 to 6 minutes, stirring once, until crisp and tender. Add soy sauce, honey, and tuna, and stir. Cook for another 3 to 6 minutes, stirring once, until the tuna is cooked as desired. Tuna can be served rare or medium-rare, or you can cook it until well done.

Per Serving: Calories: 187; Fat: 8g; Protein:17g; Fiber:2g

Pecan-Crusted Catfish

PREP: 5 MINUTES • COOK TIME: 12 MINUTES • TOTAL: 17 MINUTES • SERVES: 4

Ingredients

½ cup pecan meal
1 teaspoon fine sea salt
¼ teaspoon ground black pepper
4 (4-ounce) catfish fillets

FOR GARNISH (OPTIONAL):
Fresh oregano
Pecan halves

Instructions:

1. Spray the Innsky air fryer basket with avocado oil. Preheat the Innsky air fryer to 375°F. In a large bowl, mix the pecan meal, salt, and pepper. One at a time, dredge the catfish fillets in the mixture, coating them well. Use your hands to press the pecan meal into the fillets. Spray the fish with avocado oil and place them in the air fryer basket.
2. Cook the coated catfish for 12 minutes, or until it flakes easily and is no longer translucent in the center, flipping halfway through. Garnish with oregano sprigs and pecan halves, if desired.
   Store leftovers in an airtight container in the fridge for up to 3 days. Reheat in a preheated 350°F air fryer for 4 minutes, or until heated through.

Per serving: Calories 162; Fat 11g; Protein 17g; Total carbs 1g; Fiber 1g

Salmon Quiche

PREP: 5 MINUTES • COOK TIME: 12 MINUTES • TOTAL: 17 MINUTES • SERVES: 4

Ingredients

5 Ozs Salmon Fillet
1/2 Tbsp Lemon Juice
1/2 Cup Flour
1/4 Cup Butter, melted
2 Eggs and 1 Egg Yolk

3 Tbsps Whipped Cream
Tsps Mustard
Black Pepper to taste
Salt and Pepper
Quiche Pan

Instructions:

1. Clean and cut the salmon into small cubes. Heat the Air fryer to 375 degrees. Pour the lemon juice over the salmon cubes and allow to marinate for an hour.
   Combine a tablespoon of water with the butter, flour and yolk in a large bowl. Using your hands, knead the mixture until smooth. On a clean surface, use a rolling pin to form a circle of dough. Place this into the quiche pan, using your fingers to adhere the pastry to the edges. Whisk the cream, mustard and eggs together. Season with salt and pepper. Add the marinated salmon into the bowl and combine.
   Pour the content of the bowl into the dough lined quiche pan
2. Put the pan in the Air fryer tray and cook for 25 minutes until browned and crispy.

Lemony Tuna

PREP: 10 MINUTES • COOK TIME: 10 MINUTES • TOTAL: 20 MINUTES • SERVES: 4

Ingredients

2 (6-ounce) cans water packed plain tuna
2 teaspoons Dijon mustard
½ cup breadcrumbs
1 tablespoon fresh lime juice
2 tablespoons fresh parsley, chopped

1 egg
Dash of hot sauce
3 tablespoons canola oil
Salt and freshly ground black pepper, to taste

Instructions:

1. Drain most of the liquid from the canned tuna.
   In a bowl, add the fish, mustard, crumbs, citrus juice, parsley and hot sauce and mix till well combined. Add a little canola oil if it seems too dry. Add egg, salt and stir to combine. Make the patties from tuna mixture. Refrigerate the tuna patties for about 2 hours.
2. Preheat the Innsky air fryer to 355 degrees F. Cook for about 10-12 minutes

Friday Night Fish Fry
PREP: 10 MINUTES • COOK TIME: 10 MINUTES • TOTAL: 20 MINUTES • SERVES: 4

Ingredients
1 large egg
½ cup powdered Parmesan cheese, about 1½
1 teaspoon smoked paprika
¼ teaspoon celery salt
¼ teaspoon ground black pepper

4 (4-ounce) cod fillets
Chopped fresh oregano or parsley, for garnish (optional)
Lemon slices, for serving (optional)

Instructions:
1. Spray the Innsky air fryer basket with avocado oil. Preheat the Innsky air fryer to 400°F. Crack the egg in a shallow bowl and beat it lightly with a fork. Combine the Parmesan cheese, paprika, celery salt, and pepper in a separate shallow bowl. One at a time, dip the fillets into the egg, then dredge them in the Parmesan mixture. Using your hands, press the Parmesan onto the fillets to form a nice crust. As you finish, place the fish in the air fryer basket.
2. Cook the fish in the air fryer for 10 minutes, or until it is cooked through and flakes easily with a fork. Garnish with fresh oregano or parsley and serve with lemon slices, if desired. Store leftovers in an airtight container in the refrigerator for up to 3 days. Reheat in a preheated 400°F air fryer for 5 minutes, or until warmed through.

Per serving: Calories 164; Fat 5g; Protein 26g; Total carbs 1g; Fiber 0.2g

Bang Bang Panko Breaded Fried Shrimp
PREP: 5 MINUTES • COOK TIME: 8 MINUTES • TOTAL: 13 MINUTES • SERVES: 4

Ingredients
1 tsp. paprika
Montreal chicken seasoning
¾ C. panko bread crumbs
½ C. almond flour
1 egg white

1pound raw shrimp (peeled and deveined)
 Bang Bang Sauce:
¼ C. sweet chili sauce
2 tbsp. sriracha sauce
1/3 C. plain Greek yogurt

Instructions:
1 Ensure your air fryer is preheated to 400 degrees.
   Season all shrimp with seasonings. Add flour to one bowl, egg white in another, and breadcrumbs to a third. Dip seasoned shrimp in flour, then egg whites, and then breadcrumbs. Spray coated shrimp with olive oil and add to air fryer basket.
2 Set temperature to 400°F, and set time to 4 minutes. Cook 4 minutes, flip, and cook an additional 4 minutes. To make the sauce, mix together all sauce ingredients until smooth.

Flying Fish
PREP: 5 MINUTES • COOK TIME: 12 MINUTES • TOTAL: 17 MINUTES • SERVES: 6

Ingredients
4 Tbsp Oil
3–4 oz Breadcrumbs
1 Whisked Whole Egg in a Saucer/Soup Plate

4 Fresh Fish Fillets
Fresh Lemon (For serving)

Instructions:
1 Preheat the Innsky air fryer to 350° F. Mix the crumbs and oil until it looks nice and loose. Dip the fish in the egg and coat lightly, then move on to the crumbs. Make sure the fillet is covered evenly.
2 Cook in the Innsky air fryer basket for roughly 12 minutes – depending on the size of the fillets you are using. Serve with fresh lemon & chips to complete the duo.

Cilantro-Lime Fried Shrimp
PREP: 10 MINUTES • COOK TIME: 10 MINUTES • TOTAL: 20 MINUTES • SERVES: 4

Ingredients

1pound raw shrimp, peeled and deveined with tails on or off
½ cup chopped fresh cilantro
Juice of 1 lime
1 egg
½ cup all-purpose flour

¾ cup bread crumbs
Salt
Pepper
Cooking oil
½ cup cocktail sauce (optional

Instructions:

1   Place the shrimp in a plastic bag and add the cilantro and lime juice. Seal the bag. Shake to combine. Marinate in the refrigerator for 30 minutes.
    In a small bowl, beat the egg. In another small bowl, place the flour. Place the bread crumbs in a third small bowl, and season with salt and pepper to taste.
    Spray the Innsky air fryer basket with cooking oil. Remove the shrimp from the plastic bag. Dip each in the flour, then the egg, and then the bread crumbs.
2   Place the shrimp in the Air fryer. It is okay to stack them. Spray the shrimp with cooking oil. Cook for 4 minutes. Open the air fryer and flip the shrimp. I recommend flipping individually instead of shaking to keep the breading intact. Cook for an additional 4 minutes, or until crisp. Cool before serving. Serve with cocktail sauce if desired.

Per Serving: Calories: 254; Fat:4g; Protein:29g; Fiber:1g

Grilled Soy Salmon Fillets
PREP: 5 MINUTES • COOK TIME: 8 MINUTES • TOTAL: 13 MINUTES • SERVES: 4

Ingredients

4 salmon fillets
1/4 teaspoon ground black pepper
1/2 teaspoon cayenne pepper
1/2 teaspoon salt
1 teaspoon onion powder

1 tablespoon fresh lemon juice
1/2 cup soy sauce
1/2 cup water
1 tablespoon honey
2 tablespoons extra-virgin olive oil

Instructions:

1   Firstly, pat the salmon fillets dry using kitchen towels. Season the salmon with black pepper, cayenne pepper, salt, and onion powder.
    To make the marinade, combine together the lemon juice, soy sauce, water, honey, and olive oil. Marinate the salmon for at least 2 hours in your refrigerator.
    Arrange the fish fillets on a grill basket in your Air fryer.
2   Bake at 330 degrees for 8 to 9 minutes, or until salmon fillets are easily flaked with a fork.
    Work with batches and serve warm.

Parmesan-Crusted Shrimp over Pesto Zoodles
PREP: 10 MINUTES • COOK TIME: 7 MINUTES • TOTAL: 17 MINUTES • SERVES: 4

Ingredients
2 large eggs

3 cloves garlic, minced

2 teaspoons dried basil, divided

½ teaspoon fine sea salt

½ teaspoon ground black pepper

½ cup powdered Parmesan cheese (about 1½ ounces)

1 pound jumbo shrimp, peeled, deveined, butterflied, tails removed

PESTO:

1 packed cup fresh basil

¼ cup extra-virgin olive oil or avocado oil

¼ cup grated Parmesan cheese

¼ cup roasted, salted walnuts (omit for nut-free)

3 cloves garlic, peeled

1 tablespoon lemon juice

½ teaspoon fine sea salt

¼ teaspoon ground black pepper

2 recipes Perfect Zoodles, warm, for serving

Instructions:
1. Spray the Innsky air fryer basket with avocado oil. Preheat the Innsky air fryer to 400°F.
   In a large bowl, whisk together the eggs, garlic, 1 teaspoon of the dried basil, the salt, and the pepper. In a separate small bowl, mix together the remaining teaspoon of dried basil and the Parmesan cheese. Place the shrimp in the bowl with the egg mixture and use your hands to coat the shrimp. Roll one shrimp in the Parmesan mixture and press the coating onto the shrimp with your hands. Place the coated shrimp in the air fryer basket. Repeat with the remaining shrimp, leaving space between them in the air fryer basket. (If you're using a smaller air fryer, work in batches if necessary.)
2. Cook the shrimp in the air fryer for 7 minutes, or until cooked through and no longer translucent, flipping after 4 minutes. While the shrimp cook, make the pesto: Place all the ingredients for the pesto in a food processor and pulse until smooth, with a few rough pieces of basil. Just before serving, toss the warm zoodles with the pesto and place the shrimp on top. Store leftover shrimp and pesto zoodles in separate airtight containers in the refrigerator for up to 3 days or in the freezer for up to a month. Reheat the shrimp in a preheated 400°F air fryer for 5 minutes, or until warmed through. To reheat the pesto zoodles, place them in a casserole dish that will fit in your air fryer and cook at 350°F for 2 minutes, or until heated through.

Old Bay Crab Cakes
PREP: 10 MINUTES • COOK TIME: 20 MINUTES • TOTAL: 30 MINUTES • SERVES: 4

Ingredients
2 slices dried bread, crusts removed

Small amount of milk

1 tablespoon mayonnaise

1 tablespoon Worcestershire sauce

1 tablespoon baking powder

1 tablespoon parsley flakes

1 teaspoon Old Bay® Seasoning

1/4 teaspoon salt

1 egg

1 pound lump crabmeat

Instructions:
1. Crush your bread over a large bowl until it is broken down into small pieces. Add milk and stir until bread crumbs are moistened. Mix in mayo and Worcestershire sauce. Add remaining ingredients and mix well. Shape into 4 patties.
2. Cook at 360 degrees for 20 minutes, flip half way through.

Asian Marinated Salmon

PREP: 5 MINUTES PLUS 2 HOURS TO MARINATE • COOK TIME: 6 MINUTES • TOTAL: 2 HOURS 11 MINUTES• SERVES: 2

Ingredients

MARINADE:
¼ cup wheat-free tamari or coconut aminos
2 tablespoons lime or lemon juice
2 tablespoons sesame oil
2 tablespoons Swerve confectioners'-style sweetener, or a few drops liquid stevia
2 teaspoons grated fresh ginger
2 cloves garlic, minced
½ teaspoon ground black pepper
2 (4-ounce) salmon fillets (about 1¼ inches thick)
Sliced green onions, for garnish

SAUCE (OPTIONAL):
¼ cup beef broth
¼ cup wheat-free tamari
3 tablespoons Swerve confectioners'-style sweetener or equivalent amount of liquid or powdered sweetener
1 tablespoon tomato sauce
1 teaspoon stevia glycerite (optional)
⅛ teaspoon guar gum or xanthan gum (optional, for thickening)

Instructions:

1. Make the marinade: In a medium-sized shallow dish, stir together all the ingredients for the marinade until well combined. Place the salmon in the marinade. Cover and refrigerate for at least 2 hours or overnight. Preheat the Innsky air fryer to 400°F.
Remove the salmon fillets from the marinade and place them in the air fryer, leaving space between them.
2. Cook for 6 minutes, or until the salmon is cooked through and flakes easily with a fork.
While the salmon cooks, make the sauce, if using: Place all the sauce ingredients except the guar gum in a medium-sized bowl and stir until well combined. Taste and adjust the sweetness to your liking. While whisking slowly, add the guar gum. Allow the sauce to thicken for 3 to 5 minutes. (The sauce can be made up to 3 days ahead and stored in an airtight container in the fridge.) Drizzle the sauce over the salmon before serving. Garnish the salmon with sliced green onions before serving. Store leftovers in an airtight container in the fridge for up to 3 days. Reheat in a preheated 350°F air fryer for 3 minutes, or until heated through.

Per serving: Calories 311; Fat 18g; Protein 31g; Total carbs 9g; Fiber 1g

Louisiana Shrimp Po Boy

PREP: 10 MINUTES • COOK TIME: 10 MINUTES • TOTAL: 20 MINUTES • SERVES: 6

Ingredients

1 tsp. creole seasoning
8 slices of tomato
Lettuce leaves
¼ C. buttermilk
½ C. Louisiana Fish Fry
1 pound deveined shrimp
Remoulade sauce:

1 chopped green onion
1 tsp. hot sauce
1 tsp. Dijon mustard
½ tsp. creole seasoning
1 tsp. Worcestershire sauce
Juice of ½ a lemon
½ C. vegan mayo

Instructions:

1. To make the sauce, combine all sauce ingredients until well incorporated. Chill while you cook shrimp. Mix seasonings together and liberally season shrimp. Add buttermilk to a bowl. Dip each shrimp into milk and place in a Ziploc bag. Chill half an hour to marinate. Add fish fry to a bowl. Take shrimp from marinating bag and dip into fish fry, then add to air fryer. Ensure your air fryer is preheated to 400 degrees. Spray shrimp with olive oil.
2. Set temperature to 400°F, and set time to 5 minutes. Cook 5 minutes, flip and then cook another 5 minutes. Assemble "Keto" Po Boy by adding sauce to lettuce leaves, along with shrimp and tomato.

Per Serving: Calories: 337; Carbs:5.5; Fat: 12g; Protein:24g; Sugar:2g

BLT Crab Cakes
PREP: 10 MINUTES • COOK TIME: 19 MINUTES • TOTAL: 29 MINUTES • SERVES: 4

Ingredients
4 slices bacon
CRAB CAKES:
1 pound canned lump crabmeat, drained well
¼ cup plus 1 tablespoon powdered Parmesan cheese (or pork dust for dairy-free)
3 tablespoons mayonnaise
1 large egg
½ teaspoon dried chives
½ teaspoon dried parsley
½ teaspoon dried dill weed

¼ teaspoon garlic powder
¼ teaspoon onion powder
⅛ teaspoon ground black pepper
1 cup pork dust
FOR SERVING:
Leaves from 1 small head Boston lettuce
4 slices tomato
¼ cup mayonnaise

Instructions:
1. Spray the Innsky air fryer basket with avocado oil. Preheat the Innsky air fryer to 350°F. Place the bacon slices in the air fryer, leaving space between them, and cook for 7 to 9 minutes, until crispy. Remove the bacon and increase the heat to 400°F. Set the bacon aside.
   Make the crab cakes: Place all the crab cake ingredients except the pork dust in a large bowl and mix together with your hands until well blended. Divide the mixture into 4 equal-sized crab cakes (they should each be about 1 inch thick). Place the pork dust in a small bowl. Dredge the crab cakes in the pork dust to coat them well and use your hands to press the pork dust into the cakes.
2. Place the crab cakes in the air fryer basket, leaving space between them, and cook for 10 minutes, or until crispy on the outside. To serve, place 4 lettuce leaves on a serving platter and top each leaf with a slice of tomato, then a crab cake, then a dollop of mayo, and finally a slice of bacon.
   Store leftovers in an airtight container in the refrigerator for up to 3 days. Reheat in a preheated 350°F air fryer for 6 minutes, or until heated through.

Per serving: Calories 341; Fat 28g; Protein 22g; Total carbs 3g; Fiber 1g

Chilean Sea Bass with Olive Relish
PREP: 10 MINUTES • COOK TIME: 10 MINUTES • TOTAL: 20 MINUTES • SERVES: 2

Ingredients
Olive oil spray
2 (6-ounce) Chilean sea bass fillets or other firm-fleshed white fish
3 tablespoons extra-virgin olive oil
½ teaspoon ground cumin

½ teaspoon kosher salt
½ teaspoon black pepper
⅓ cup pitted green olives, diced
¼ cup finely diced onion
1 teaspoon chopped capers

Instructions:
1. Spray the Innsky air fryer basket with the olive oil spray. Drizzle the fillets with the olive oil and sprinkle with the cumin, salt, and pepper. Place the fish in the air fryer basket.
2. Set the Innsky air fryer to 325°F for 10 minutes, or until the fish flakes easily with a fork.
   Meanwhile, in a small bowl, stir together the olives, onion, and capers.
   Serve the fish topped with the relish.

Mouthwatering Cod over Creamy Leek Noodles
PREP: 10 MINUTES • COOK TIME: 24 MINUTES • TOTAL: 34 MINUTES • SERVES: 4

Ingredients

1 small leek, sliced into long thin noodles (about 2 cups)
½ cup heavy cream
2 cloves garlic, minced
1 teaspoon fine sea salt, divided
4 (4-ounce) cod fillets (about 1 inch thick)
½ teaspoon ground black pepper

COATING:
¼ cup grated Parmesan cheese
2 tablespoons mayonnaise
2 tablespoons unsalted butter, softened
1 tablespoon chopped fresh thyme, or ½ teaspoon dried thyme leaves, plus more for garnish

Instructions:

1. Preheat the Innsky air fryer to 350°F. Place the leek noodles in a 6-inch casserole dish or a pan that will fit in your air fryer. In a small bowl, stir together the cream, garlic, and ½ teaspoon of the salt.

2. Pour the mixture over the leeks and cook in the air fryer for 10 minutes, or until the leeks are very tender. Pat the fish dry and season with the remaining ½ teaspoon of salt and the pepper. When the leeks are ready, open the air fryer and place the fish fillets on top of the leeks. Cook for 8 to 10 minutes, until the fish flakes easily with a fork. While the fish cooks, make the coating: In a small bowl, combine the Parmesan, mayo, butter, and thyme. When the fish is ready, remove it from the air fryer and increase the heat to 425°F (or as high as your air fryer can go). Spread the fillets with a ½-inch-thick to ¾-inch-thick layer of the coating. Place the fish back in the air fryer and cook for 3 to 4 minutes, until the coating browns. Garnish with fresh or dried thyme, if desired. Store leftovers in an airtight container in the refrigerator for up to 3 days. Reheat in a casserole dish in a preheated 350°F air fryer for 6 minutes, or until heated through.

Spicy Popcorn Shrimp
PREP: 10 MINUTES • COOK TIME: 9 MINUTES • TOTAL: 19 MINUTES • SERVES: 4

Ingredients

4 large egg yolks
1 teaspoon prepared yellow mustard
1 pound small shrimp, peeled, deveined, and tails removed
½ cup finely shredded Gouda or Parmesan cheese
½ cup pork dust

1 tablespoon Cajun seasoning
FOR SERVING/GARNISH (OPTIONAL):
Prepared yellow mustard
Ranch Dressing
Tomato sauce
Sprig of fresh parsley

Instructions:

1. Spray the Innsky air fryer basket with avocado oil. Preheat the Innsky air fryer to 400°F. Place the egg yolks in a large bowl, add the mustard, and whisk until well combined. Add the shrimp and stir well to coat.
In a medium-sized bowl, mix together the cheese, pork dust, and Cajun seasoning until well combined. One at a time, roll the coated shrimp in the pork dust mixture and use your hands to press it onto the shrimp. Spray the coated shrimp with avocado oil and place them in the air fryer basket, leaving space between them.

2. Cook the shrimp in the air fryer for 9 minutes, or until cooked through and no longer translucent, flipping after 4 minutes. Serve with your dipping sauces of choice and garnish with a sprig of fresh parsley. Store leftovers in an airtight container in the refrigerator for up to 3 days. Reheat in a preheated 400°F air fryer for 5 minutes, or until warmed through.

Per serving: Calories 199; Fat 9g; Protein 27g; Total carbs 1g; Fiber 0g

Pistachio-Crusted Lemon-Garlic Salmon
PREP: 5 MINUTES • COOK TIME: 20 MINUTES • TOTAL: 25 MINUTES • SERVES: 6

Ingredients

4 medium-sized salmon filets

2 raw eggs

3 ounces of melted butter

1 clove of garlic, peeled and finely minced

1 large-sized lemon

1 teaspoon of salt

1 tablespoon of parsley, rinsed, patted dry and chopped

1 teaspoon of dill, rinsed, patted dry and chopped

½ cup of pistachio nuts, shelled and coarsely crushed

Instructions:

1   Cover the basket of the Air fryer with a lining of tin foil, leaving the edges uncovered to allow air to circulate through the basket. Preheat the Innsky air fryer to 350 degrees. In a mixing bowl, beat the eggs until fluffy and until the yolks and whites are fully combined. Add the melted butter, the juice of the lemon, the minced garlic, the parsley and the dill to the beaten eggs, and stir thoroughly.

   One by one, dunk the salmon filets into the wet mixture, then roll them in the crushed pistachios, coating completely. Place the coated salmon fillets in the Air fryer basket.

2   Set the Innsky air fryer timer for 10 minutes. When the air fryer shuts off, after 10 minutes, the salmon will be partly cooked and the crust beginning to crisp. Using tongs, turn each of the fish filets over. ReSet the Innsky air fryer to 350 degrees for another 10 minutes. After 10 minutes, when the air fryer shuts off, the salmon will be perfectly cooked and the pistachio crust will be toasted and crispy. Using tongs, remove from the Air fryer and serve.

Breaded Shrimp Tacos
PREP: 10 MINUTES • COOK TIME: 9 MINUTES • TOTAL: 19 MINUTES • SERVES: 8 TACOS

Ingredients

2 large eggs

1 teaspoon prepared yellow mustard

1 pound small shrimp, peeled, deveined, and tails removed

½ cup finely shredded Gouda or Parmesan cheese

½ cup pork dust

FOR SERVING:

8 large Boston lettuce leaves

¼ cup pico de gallo

¼ cup shredded purple cabbage

1 lemon, sliced

Guacamole (optional)

Instructions:

1.   Preheat the Innsky air fryer to 400°F.

   Crack the eggs into a large bowl, add the mustard, and whisk until well combined. Add the shrimp and stir well to coat. In a medium-sized bowl, mix together the cheese and pork dust until well combined. One at a time, roll the coated shrimp in the pork dust mixture and use your hands to press it onto each shrimp. Spray the coated shrimp with avocado oil and place them in the air fryer basket, leaving space between them.

2.   Cook the shrimp for 9 minutes, or until cooked through and no longer translucent, flipping after 4 minutes. To serve, place a lettuce leaf on a serving plate, place several shrimp on top, and top with 1½ teaspoons each of pico de gallo and purple cabbage. Squeeze some lemon juice on top and serve with guacamole, if desired.

   Store leftover shrimp in an airtight container in the refrigerator for up to 3 days. Reheat in a preheated 400°F air fryer for 5 minutes, or until warmed through.

Per serving: Calories 194; Fat 8g; Protein 28g; Total carbs 3g; Fiber 0.5g

Air Fryer Salmon Patties
PREP: 8 MINUTES • COOK TIME: 7 MINUTES • TOTAL: 15 MINUTES • SERVES: 4

Ingredients

1 tbsp. olive oil

1 tbsp. ghee

¼ tsp. salt

1/8 tsp. pepper

1 egg

1 C. almond flour

1 can wild Alaskan pink salmon

Instructions:

1. Drain can of salmon into a bowl and keep liquid. Discard skin and bones. Add salt, pepper, and egg to salmon, mixing well with hands to incorporate. Make patties. Dredge in flour and remaining egg. If it seems dry, spoon reserved salmon liquid from the can onto patties.
2. Add patties to the Air fryer. Cook 7 minutes at 378 degrees till golden, making sure to flip once during cooking process.

Green Curry Shrimp
PREP: 15 MINUTES PLUS 15 MINUTES TO MARINATE • COOK TIME: 5 MINUTES • TOTAL: 35 MINUTES • SERVES: 4

Ingredients

1 to 2 tablespoons Thai green curry paste

2 tablespoons coconut oil, melted

1 tablespoon half-and-half or coconut milk

1 teaspoon fish sauce

1 teaspoon soy sauce

1 teaspoon minced fresh ginger

1 clove garlic, minced

1 pound jumbo raw shrimp (21 to 25 count), peeled and deveined

¼ cup chopped fresh Thai basil or sweet basil

¼ cup chopped fresh cilantro

Instructions:

1 In a 6 × 3-inch round heatproof pan, combine the curry paste, coconut oil, half-and-half, fish sauce, soy sauce, ginger, and garlic. Whisk until well combined. Add the shrimp and toss until well coated. Marinate at room temperature for 15 to 30 minutes. Place the pan in the air fryer basket.
2 Set the Innsky air fryer to 400°F for 5 minutes, stirring halfway through the cooking time. Transfer the shrimp to a serving bowl or platter. Garnish with the basil and cilantro.

Chinese Ginger-Scallion Fish
PREP: 15 MINUTES • COOK TIME: 15 MINUTES • TOTAL: 30 MINUTES • SERVES: 2

Ingredients

For the Bean Sauce

2 tablespoons soy sauce

1 tablespoon rice wine

1 tablespoon **doubanjiang** (Chinese black bean paste)

1 teaspoon minced fresh ginger

1 clove garlic, minced

For the Vegetables and Fish

1 tablespoon peanut oil

¼ cup julienned green onions (white and green parts)

¼ cup chopped fresh cilantro

2 tablespoons julienned fresh ginger

2 (6-ounce) white fish fillets, such as tilapia

Instructions:

1. For the sauce: In a small bowl, combine all the ingredients and stir until well combined; set aside. For the vegetables and fish: In a medium bowl, combine the peanut oil, green onions, cilantro, and ginger. Toss to combine. Cut two squares of parchment large enough to hold one fillet and half of the vegetables. Place one fillet on each parchment square, top with the vegetables, and pour over the sauce. Fold over the parchment paper and crimp the sides in small, tight folds to hold the fish, vegetables, and sauce securely inside the packet. Place the packets in a single layer in the air fryer basket. Set fryer to 350°F for 15 minutes. Transfer each packet to a dinner plate. Cut open with scissors just before serving.

Bang Bang Shrimp

PREP: 15 MINUTES • COOK TIME: 14 MINUTES • TOTAL: 29 MINUTES • SERVES: 4

Ingredients

For the Sauce
½ cup mayonnaise
¼ cup sweet chili sauce
2 to 4 tablespoons sriracha
1 teaspoon minced fresh ginger
For the Shrimp

1 pound jumbo raw shrimp (21 to 25 count), peeled and deveined
2 tablespoons cornstarch or rice flour
½ teaspoon kosher salt
Vegetable oil spray

Instructions:

1. For the sauce: In a large bowl, combine the mayonnaise, chili sauce, sriracha, and ginger. Stir until well combined. Remove half of the sauce to serve as a dipping sauce.
   For the shrimp: Place the shrimp in a medium bowl. Sprinkle the cornstarch and salt over the shrimp and toss until well coated. Place the shrimp in the Innsky air fryer basket in a single layer. (If they won't fit in a single layer, set a rack or trivet on top of the bottom layer of shrimp and place the rest of the shrimp on the rack.) Spray generously with vegetable oil spray.
2. Set the Innsky air fryer to 350°F for 10 minutes, turning and spraying with additional oil spray halfway through the cooking time. Remove the shrimp and toss in the bowl with half of the sauce. Place the shrimp back in the air fryer basket. Set the Innsky air fryer to 350°F for an additional 4 to 5 minutes, or until the sauce has formed a glaze.
   Serve the hot shrimp with the reserved sauce for dipping.

Fried Calamari

PREP: 8 MINUTES • COOK TIME: 7 MINUTES • TOTAL: 15 MINUTES • SERVES: 6-8

Ingredients

½ tsp. salt
½ tsp. Old Bay seasoning
1/3 C. plain cornmeal
½ C. semolina flour

½ C. almond flour
5-6 C. olive oil
1 ½ pounds baby squid

Instructions:

1. Rinse squid in cold water and slice tentacles, keeping just ¼-inch of the hood in one piece. Combine 1-2 pinches of pepper, salt, Old Bay seasoning, cornmeal, and both flours together. Dredge squid pieces into flour mixture and place into the air fryer.
2. Spray liberally with olive oil. Cook 15 minutes at 345 degrees till coating turns a golden brown.

Soy and Ginger Shrimp

PREP: 8 MINUTES • COOK TIME: 10 MINUTES • TOTAL: 15 MINUTES • SERVES: 4

Ingredients

2 tablespoons olive oil
2 tablespoons scallions, finely chopped
2 cloves garlic, chopped
1 teaspoon fresh ginger, grated
1 tablespoon dry white wine

1 tablespoon balsamic vinegar
1/4 cup soy sauce
1 tablespoon sugar
1 pound shrimp
Salt and ground black pepper, to taste

Instructions:

1. To make the marinade, warm the oil in a saucepan; cook all ingredients, except the shrimp, salt, and black pepper. Now, let it cool. Marinate the shrimp, covered, at least an hour, in the refrigerator.
2. After that, bake the shrimp at 350 degrees F for 8 to 10 minutes turning once or twice. Season prepared shrimp with salt and black pepper and serve right away.

Cajun Fried Shrimp with Remoulade

PREP: 30 MINUTES PLUS 15 MINUTES TO CHILL• COOK TIME: 8 MINUTES • TOTAL: 53 MINUTES• SERVES: 4

Ingredients

For the Remoulade
½ cup mayonnaise
1 green onion, finely chopped
1 clove garlic, minced
1 tablespoon sweet pickle relish
2 tablespoons Creole mustard
2 teaspoons fresh lemon juice
½ teaspoon hot pepper sauce
½ teaspoon Worcestershire sauce
¼ teaspoon smoked paprika

¼ teaspoon kosher salt
For the Shrimp
1½ cups buttermilk
1 large egg
3 teaspoons salt-free Cajun seasoning
1 pound jumbo raw shrimp (21 to 25 count), peeled and deveined
2 cups finely ground cornmeal
Kosher salt and black pepper
Vegetable oil spray

Instructions:

1. For the remoulade: In a small bowl, stir together all the ingredients until well combined. Cover the sauce and chill until serving time.
   For the shrimp: In a large bowl, whisk together the buttermilk, egg, and 1 teaspoon of the Cajun seasoning. Add the shrimp and toss gently to combine. Refrigerate for at least 15 minutes, or up to 1 hour. Meanwhile, in a shallow dish, whisk together the remaining 2 teaspoons Cajun seasoning, cornmeal, and salt and pepper to taste. Spray the Innsky air fryer basket with the vegetable oil spray. Dredge the shrimp in the cornmeal mixture until well coated. Shake off any excess and arrange the shrimp in the air fryer basket. Spray with oil spray.
2. Set the Innsky air fryer to 350°F for 8 minutes, carefully turning and spraying the shrimp with the oil spray halfway through the cooking time. Serve the shrimp with the remoulade.

Salmon Noodles

PREP: 5 MINUTES • COOK TIME: 16 MINUTES • TOTAL: 21 MINUTES • SERVES: 4

Ingredients

1 Salmon Fillet
1 Tbsp Teriyaki Marinade
3 ½ Ozs Soba Noodles, cooked and drained
10 Ozs Firm Tofu

7 Ozs Mixed Salad
1 Cup Broccoli
Olive Oil
Salt and Pepper to taste

Instructions:

1 Season the salmon with salt and pepper to taste, then coat with the teriyaki marinate. Set aside for 15 minutes
2 Preheat the Innsky air fryer at 350 degrees, then cook the salmon for 8 minutes. Whilst the Air fryer is cooking the salmon, start slicing the tofu into small cubes. Next, slice the broccoli into smaller chunks. Drizzle with olive oil.
   Once the salmon is cooked, put the broccoli and tofu into the Air fryer tray for 8 minutes. Plate the salmon and broccoli tofu mixture over the soba noodles. Add the mixed salad to the side and serve

Scallops and Spring Veggies
PREP: 10 MINUTES • COOK TIME: 8 MINUTES • TOTAL: 18 MINUTES • SERVES: 4

Ingredients

½ pound asparagus, ends trimmed, cut into 2-inch pieces
1 cup sugar snap peas
1pound sea scallops
1 tablespoon lemon juice

2 teaspoons olive oil
½ teaspoon dried thyme
Pinch salt
Freshly ground black pepper

Instructions:

1   Place the asparagus and sugar snap peas in the Air fryer basket.
2   Cook for 2 to 3 minutes or until the vegetables are just starting to get tender.
    Meanwhile, check the scallops for a small muscle attached to the side, and pull it off and discard.
    In a medium bowl, toss the scallops with the lemon juice, olive oil, thyme, salt, and pepper. Place into the Innsky air fryer basket on top of the vegetables.
3   Steam for 5 to 7 minutes, tossing the basket once during cooking time, until the scallops are just firm when tested with your finger and are opaque in the center, and the vegetables are tender. Serve immediately.

Per Serving: Calories: 162; Carbs:10g; Fat: 4g; Protein:22g; Fiber:3g

Chinese Ginger-Scallion Fish
PREP: 15 MINUTES • COOK TIME: 15 MINUTES • TOTAL: 30 MINUTES • SERVES: 2

Ingredients

For the Bean Sauce
2 tablespoons soy sauce
1 tablespoon rice wine
1 tablespoon **doubanjiang** (Chinese black bean paste)
1 teaspoon minced fresh ginger
1 clove garlic, minced

For the Vegetables and Fish
1 tablespoon peanut oil
¼ cup julienned green onions (white and green parts)
¼ cup chopped fresh cilantro
2 tablespoons julienned fresh ginger
2 (6-ounce) white fish fillets, such as tilapia

Instructions:

1   For the sauce: In a small bowl, combine all the ingredients and stir until well combined; set aside. For the vegetables and fish: In a medium bowl, combine the peanut oil, green onions, cilantro, and ginger. Toss to combine.
    Cut two squares of parchment large enough to hold one fillet and half of the vegetables. Place one fillet on each parchment square, top with the vegetables, and pour over the sauce. Fold over the parchment paper and crimp the sides in small, tight folds to hold the fish, vegetables, and sauce securely inside the packet. Place the packets in a single layer in the air fryer basket.
2   Set fryer to 350°F for 15 minutes. Transfer each packet to a dinner plate. Cut open with scissors just before serving.

Beer-Battered Fish and Chips

PREP: 5 MINUTES • COOK TIME: 30 MINUTES • TOTAL: 35 MINUTES • SERVES: 4

Ingredients

2 eggs

1 cup malty beer, such as Pabst Blue Ribbon

1 cup all-purpose flour

½ cup cornstarch

1 teaspoon garlic powder

Salt

Pepper

Cooking oil

(4-ounce) cod fillets

Instructions:

1   In a medium bowl, beat the eggs with the beer. In another medium bowl, combine the flour and cornstarch, and season with the garlic powder and salt and pepper to taste. Spray the Innsky air fryer basket with cooking oil. Dip each cod fillet in the flour and cornstarch mixture and then in the egg and beer mixture. Dip the cod in the flour and cornstarch a second time.

2   Place the cod in the Air fryer. Do not stack. Cook in batches. Spray with cooking oil. Cook for 8 minutes. Open the Air fryer and flip the cod. Cook for an additional 7 minutes. Remove the cooked cod from the Air fryer, then repeat steps 4 and 5 for the remaining fillets. Serve with prepared air fried frozen fries. Frozen fries will need to be cooked for 18 to 20 minutes at 400ºF. Cool before serving.

Per Serving: Calories: 325; Carbs:41; Fat: 4g; Protein:26g; Fiber:1g

Tuna Stuffed Potatoes

PREP: 5 MINUTES • COOK TIME: 30 MINUTES • TOTAL: 35 MINUTES • SERVES: 4

Ingredients

4 starchy potatoes

½ tablespoon olive oil

1 (6-ounce) can tuna, drained

2 tablespoons plain Greek yogurt

1 teaspoon red chili powder

Salt and freshly ground black pepper, to taste

1 scallion, chopped and divided

1 tablespoon capers

Instructions:

1   In a large bowl of water, soak the potatoes for about 30 minutes. Drain well and pat dry with paper towel. Preheat the Innsky air fryer to 355 degrees F. Place the potatoes in a fryer basket.

2   Cook for about 30 minutes. Meanwhile in a bowl, add tuna, yogurt, red chili powder, salt, black pepper and half of scallion and with a potato masher, mash the mixture completely. Remove the potatoes from the Air fryer and place onto a smooth surface. Carefully, cut each potato from top side lengthwise.

With your fingers, press the open side of potato halves slightly. Stuff the potato open portion with tuna mixture evenly.

Sprinkle with the capers and remaining scallion. Serve immediately.

One-Pot Shrimp Fried Rice
PREP: 10 MINUTES • COOK TIME: 25 MINUTES • TOTAL: 35 MINUTES • SERVES: 4

Ingredients

For the Shrimp
1 teaspoon cornstarch
½ teaspoon kosher salt
¼ teaspoon black pepper
1 pound jumbo raw shrimp (21 to 25 count), peeled and deveined
For the Rice
2 cups cold cooked rice
1 cup frozen peas and carrots, thawed

¼ cup chopped green onions (white and green parts)
3 tablespoons toasted sesame oil
1 tablespoon soy sauce
½ teaspoon kosher salt
1 teaspoon black pepper
For the Eggs
2 large eggs, beaten
¼ teaspoon kosher salt
¼ teaspoon black pepper

Instructions:

1   For the shrimp: In a small bowl, whisk together the cornstarch, salt, and pepper until well combined. Place the shrimp in a large bowl and sprinkle the seasoned cornstarch over. Toss until well coated; set aside. For the rice: In a 6 × 3-inch round heatproof pan, combine the rice, peas and carrots, green onions, sesame oil, soy sauce, salt, and pepper. Toss and stir until well combined. Place the pan in the air fryer basket. Set the Innsky air fryer to 350°F for 15 minutes, stirring and tossing the rice halfway through the cooking time. Place the shrimp on top of the rice.

2   Set the Innsky air fryer to 350°F for 5 minutes. Meanwhile, for the eggs: In a medium bowl, beat the eggs with the salt and pepper. Open the air fryer and pour the eggs over the shrimp and rice mixture. Set the Innsky air fryer to 350°F for 5 minutes. Remove the pan from the air fryer. Stir to break up the rice and mix in the eggs and shrimp.

Panko-Crusted Tilapia
PREP: 5 MINUTES • COOK TIME: 10 MINUTES • TOTAL: 15 MINUTES • SERVES: 3

Ingredients

2 tsp. Italian seasoning
2 tsp. lemon pepper
1/3 C. panko breadcrumbs
1/3 C. egg whites

1/3 C. almond flour
3 tilapia fillets
Olive oil

Instructions:

1. Place panko, egg whites, and flour into separate bowls. Mix lemon pepper and Italian seasoning in with breadcrumbs.
    Pat tilapia fillets dry. Dredge in flour, then egg, then breadcrumb mixture.
2. Add to the Innsky air fryer basket and spray lightly with olive oil.
    Cook 10-11 minutes at 400 degrees, making sure to flip halfway through cooking.

Per Serving: Calories: 256; Fat: 9g; Protein:39g; Sugar:5g

Salmon Croquettes
PREP: 5 MINUTES • COOK TIME: 10 MINUTES • TOTAL: 15 MINUTES • *SERVES: 6-8*

Ingredients
Panko breadcrumbs

Almond flour

2 egg whites

2 tbsp. chopped chives

2 tbsp. minced garlic cloves

½ C. chopped onion

2/3 C. grated carrots

1 pound chopped salmon fillet

Instructions:
1. Mix together all ingredients minus breadcrumbs, flour, and egg whites. Shape mixture into balls. Then coat them in flour, then egg, and then breadcrumbs. Drizzle with olive oil.
2. Add coated salmon balls to air fryer and cook 6 minutes at 350 degrees. Shake and cook an additional 4 minutes until golden in color.

per serving: calories: 503; carbs:61g; fat: 9g; protein:5g; sugar:4g

Potato Crusted Salmon
PREP: 10 MINUTES • COOK TIME: 15 MINUTES • TOTAL: 25 MINUTES • SERVES: 4

Ingredients
1 pound salmon, swordfish or arctic char fillets, 3/4 inch thick

1 egg white

2 tablespoons water

1/3 cup dry instant mashed potatoes

2 teaspoons cornstarch

1 teaspoon paprika

1 teaspoon lemon pepper seasoning

Instructions:
1. Remove and skin from the fish and cut it into 4 serving pieces Mix together the egg white and water. Mix together all of the dry ingredients. Dip the filets into the egg white mixture then press into the potato mix to coat evenly.
2. In your air fryer, cook at 360 degrees for 15 minutes, flip the filets halfway through.

Per Serving: Calories:176; Fat: 7g; Protein:23g; :5g

Pesto Fish Pie
PREP: 15 MINUTES • COOK TIME: 15 MINUTES • TOTAL: 30 MINUTES • SERVES: 4

Ingredients
2 tablespoons prepared pesto

¼ cup half-and-half

¼ cup grated Parmesan cheese

1 teaspoon kosher salt

1 teaspoon black pepper

Vegetable oil spray

1 (10-ounce) package frozen chopped spinach

1 pound firm white fish, cut into 2-inch chunks

½ cup cherry tomatoes, quartered

All-purpose flour

½ sheet frozen puff pastry (from a 17.3-ounce package), thaw

Instructions:
1. In a small bowl, combine the pesto, half-and-half, Parmesan, salt, and pepper. Stir until well combined; set aside. Spray a 7 × 3-inch round heatproof pan with vegetable oil spray. Arrange the spinach evenly across the bottom of the pan. Top with the fish and tomatoes. Pour the pesto mixture evenly over everything.
On a lightly floured surface, roll the puff pastry sheet into a circle. Place the pastry on top of the pan and tuck it in around the edges of the pan.
2. Place the pan in the air fryer basket. Set the Innsky air fryer to 400°F for 15 minutes, or until the pastry is well browned. Let stand 5 minutes before serving.

Snapper Scampi

PREP: 5 MINUTES • COOK TIME: 10 MINUTES • TOTAL: 15 MINUTES • *SERVES: 4*

Ingredients

4 (6-ounce) skinless snapper or arctic char fillets
1 tablespoon olive oil
3 tablespoons lemon juice, divided
½ teaspoon dried basil

Pinch salt
Freshly ground black pepper
2 tablespoons butter
cloves garlic, minced

Instructions:

1   Rub the fish fillets with olive oil and 1 tablespoon of the lemon juice. Sprinkle with the basil, salt, and pepper, and place in the Air fryer basket.
2   Grill the fish for 7 to 8 minutes or until the fish just flakes when tested with a fork. Remove the fish from the basket and put on a serving plate. Cover to keep warm. In a 6-by-6-by-2-inch pan, combine the butter, remaining 2 tablespoons lemon juice, and garlic. Cook in the Air fryer for 1 to 2 minutes or until the garlic is sizzling. Pour this mixture over the fish and serve.

Crispy Cheesy Fish Fingers

PREP: 10 MINUTES • COOK TIME: 20 MINUTES • TOTAL: 30 MINUTES • SERVES: 4

Ingredients

Large cod fish filet cut into 1 ½-inch strips
2 raw eggs
½ cup of breadcrumbs

2 tablespoons of shredded or powdered parmesan cheese
1 tablespoons of shredded cheddar cheese
Pinch of salt and pepper

Instructions:

1   Cover the basket of the Air fryer with a lining of tin foil, leaving the edges uncovered to allow air to circulate through the basket. Preheat the Innsky air fryer to 350 degrees. In a large mixing bowl, beat the eggs until fluffy and until the yolks and whites are fully combined. Dunk all the fish strips in the beaten eggs, fully submerging. In a separate mixing bowl, combine the bread crumbs with the parmesan, cheddar, and salt and pepper, until evenly mixed. One by one, coat the egg-covered fish strips in the mixed dry ingredients so that they're fully covered, and place on the foil-lined Air fryer basket.
2   Set the air-fryer timer to 20 minutes.
Halfway through the cooking time, shake the handle of the air-fryer so that the breaded fish jostles inside and fry-coverage is even. After 20 minutes, when the fryer shuts off, the fish strips will be perfectly cooked and their breaded crust golden-brown and delicious! Using tongs, remove from the air fryer and set on a serving dish to cool.

Air Fryer Fish Tacos

PREP: 5 MINUTES • COOK TIME: 15 MINUTES • TOTAL: 20 MINUTES • *SERVES: 4*

Ingredients

1pound cod
1 tbsp. cumin
½ tbsp. chili powder
1 ½ C. almond flour

1 ½ C. coconut flour
10 ounces Mexican beer
2 eggs

Instructions:

1. Whisk beer and eggs together. Whisk flours, pepper, salt, cumin, and chili powder together. Slice cod into large pieces and coat in egg mixture then flour mixture.
2. Spray bottom of your Innsky air fryer basket with olive oil and add coated codpieces. Cook 15 minutes at 375 degrees.
Serve on lettuce leaves topped with homemade salsa.
per serving: calories: 178; carbs:61g; fat:10g; protein:19g; sugar

South Indian Fried Fish

PREP: 10 MINUTES PLUS 10 MINUTES TO MARINATE • COOK TIME: 8 MINUTES • TOTAL: 28 MINUTES • SERVES: 4

Ingredients

2 tablespoons olive oil
2 tablespoons fresh lime or lemon juice
1 teaspoon minced fresh ginger
1 clove garlic, minced
1 teaspoon ground turmeric

½ teaspoon kosher salt
¼ to ½ teaspoon cayenne pepper
1 pound tilapia fillets (2 to 3 fillets)
Olive oil spray
Lime or lemon wedges (optional)

Instructions:

1 In a large bowl, combine the oil, lime juice, ginger, garlic, turmeric, salt, and cayenne. Stir until well combined; set aside. Cut each tilapia fillet into three or four equal-size pieces. Add the fish to the bowl and gently mix until all of the fish is coated in the marinade. Marinate for 10 to 15 minutes at room temperature. Spray the Innsky air fryer basket with olive oil spray. Place the fish in the basket and spray the fish.

2 Set the Innsky air fryer to 325°F for 3 minutes to partially cook the fish. Set the Innsky air fryer to 400°F for 5 minutes to finish cooking and crisp up the fish. Carefully remove the fish from the basket. Serve hot, with lemon wedges if desired.

Tandoori Shrimp

PREP: 10 MINUTES PLUS 15 MINUTES TO MARINATE • COOK TIME: 6 MINUTES • TOTAL: 31 MINUTES • SERVES: 4

Ingredients

1 pound jumbo raw shrimp (21 to 25 count), peeled and deveined
1 tablespoon minced fresh ginger
3 cloves garlic, minced
¼ cup chopped fresh cilantro or parsley, plus more for garnish
1 teaspoon ground turmeric

1 teaspoon Garam Masala
1 teaspoon smoked paprika
1 teaspoon kosher salt
½ to 1 teaspoon cayenne pepper
2 tablespoons olive oil or melted ghee
2 teaspoons fresh lemon juice

Instructions:

1 In a large bowl, combine the shrimp, ginger, garlic, cilantro, turmeric, garam masala, paprika, salt, and cayenne. Toss well to coat. Add the oil or ghee and toss again. Marinate at room temperature for 15 minutes, or cover and refrigerate for up to 8 hours.
Place the shrimp in a single layer in the air fryer basket.

2 Set the Innsky air fryer to 325°F for 6 minutes. Transfer the shrimp to a serving platter. Cover and let the shrimp finish cooking in the residual heat, about 5 minutes.
Sprinkle the shrimp with the lemon juice and toss to coat. Garnish with additional cilantro and serve.

Bacon Wrapped Scallops

PREP: 5 MINUTES • COOK TIME: 5 MINUTES • TOTAL: 10 MINUTES • *SERVES: 4*

Ingredients

1 tsp. paprika
1 tsp. lemon pepper

5 slices of center-cut bacon
20 raw sea scallops

Instructions:

1. Rinse and drain scallops, placing on paper towels to soak up excess moisture. Cut slices of bacon into 4 pieces. Wrap each scallop with a piece of bacon, using toothpicks to secure. Sprinkle wrapped scallops with paprika and lemon pepper.

2. Spray Innsky air fryer basket with olive oil and add scallops.
Cook 5-6 minutes at 400 degrees, making sure to flip halfway through.

per serving: calories: 389; carbs:63g; fat:17g; protein:21g; sugar:1g

Scallops Gratiné with Parmesan
PREP: 10 MINUTES • COOK TIME: 9 MINUTES • TOTAL: 19 MINUTES • SERVES: 2

Ingredients

For the Scallops
½ cup half-and-half
½ cup grated Parmesan cheese
¼ cup thinly sliced green onions
¼ cup chopped fresh parsley
3 cloves garlic, minced
½ teaspoon kosher salt
½ teaspoon black pepper

1 pound sea scallops
For the Topping
¼ cup crushed pork rinds or panko bread crumbs
¼ cup grated Parmesan cheese
Vegetable oil spray
For Serving
Lemon wedges
Crusty French bread (optional)

Instructions:

1. For the scallops: In a 6 × 2-inch round heatproof pan, combine the half-and-half, cheese, green onions, parsley, garlic, salt, and pepper. Stir in the scallops. For the topping: In a small bowl, combine the pork rinds or bread crumbs and cheese. Sprinkle evenly over the scallops. Spray the topping with vegetable oil spray. Place the pan in the air fryer basket.
2. Set the Innsky air fryer to 325°F for 6 minutes. Set the Innsky air fryer to 400°F for 3 minutes until the topping has browned. To serve: Squeeze the lemon wedges over the gratin and serve with crusty French bread, if desired.

Tha Fish Cakes With Mango Relish
PREP: 5 MINUTES • COOK TIME: 10 MINUTES • TOTAL: 15 MINUTES • *SERVES: 4*

Ingredients

1 lb White Fish Fillets
3 Tbsps Ground Coconut
1 Ripened Mango
½ Tsps Chili Paste
Tbsps Fresh Parsley

1 Green Onion
1 Lime
1 Tsp Salt
1 Egg

Instructions:

1. To make the relish, peel and dice the mango into cubes. Combine with a half teaspoon of chili paste, a tablespoon of parsley, and the zest and juice of half a lime. In a food processor, pulse the fish until it forms a smooth texture. Place into a bowl and add the salt, egg, chopped green onion, parsley, two tablespoons of the coconut, and the remainder of the chili paste and lime zest and juice. Combine well
Portion the mixture into 10 equal balls and flatten them into small patties. Pour the reserved tablespoon of coconut onto a dish and roll the patties over to coat. Preheat the Innsky air fryer to 390 degrees
2. Place the fish cakes into the Air fryer and cook for 8 minutes. They should be crisp and lightly browned when ready. Serve hot with mango relish

Firecracker Shrimp

PREP: 10 MINUTES • COOK TIME: 8 MINUTES • TOTAL: 18 MINUTES • *SERVES: 4*

Ingredients

For the shrimp
1 pound raw shrimp, peeled and deveined
Salt
Pepper
1 egg
½ cup all-purpose flour

¾ cup panko bread crumbs
Cooking oil
For the firecracker sauce
⅓ cup sour cream
2 tablespoons Sriracha
¼ cup sweet chili sauce

Instructions:

1   Season the shrimp with salt and pepper to taste. In a small bowl, beat the egg. In another small bowl, place the flour. In a third small bowl, add the panko bread crumbs.
Spray the Innsky air fryer basket with cooking oil. Dip the shrimp in the flour, then the egg, and then the bread crumbs. Place the shrimp in the Air fryer basket. It is okay to stack them. Spray the shrimp with cooking oil.

2   Cook for 4 minutes. Open the Air fryer and flip the shrimp. I recommend flipping individually instead of shaking to keep the breading intact. Cook for an additional 4 minutes or until crisp.
While the shrimp is cooking, make the firecracker sauce: In a small bowl, combine the sour cream, Sriracha, and sweet chili sauce. Mix well. Serve with the shrimp.

per serving: calories: 266; carbs:23g; fat:6g; protein:27g; fiber:1g

Sesame Seeds Coated Fish

PREP: 10 MINUTES • COOK TIME: 8 MINUTES • TOTAL: 18 MINUTES • *SERVES:5*

Ingredients

3 tablespoons plain flour
2 eggs
½ cup sesame seeds, toasted
½ cup breadcrumbs
1/8 teaspoon dried rosemary, crushed

Pinch of salt
Pinch of black pepper
3 tablespoons olive oil
5 frozen fish fillets (white fish of your choice)

Instructions:

1   In a shallow dish, place flour. In a second shallow dish, beat the eggs. In a third shallow dish, add remaining ingredients except fish fillets and mix till a crumbly mixture forms. Coat the fillets with flour and shake off the excess flour. Next, dip the fillets in egg. Then coat the fillets with sesame seeds mixture generously.
Preheat the Innsky air fryer to 390 degrees F.

2   Line an Innsky air fryer basket with a piece of foil. Arrange the fillets into prepared basket. Cook for about 14 minutes, flipping once after 10 minutes.

Flaky Fish Quesadilla

PREP: 10 MINUTES • COOK TIME: 12 MINUTES • TOTAL: 22 MINUTES • *SERVES: 4*

Ingredients

Two 6-inch corn or flour tortilla shells
1 medium-sized tilapia fillet, approximately 4 ounces
½ medium-sized lemon, sliced
½ an avocado, peeled, pitted and sliced
1 clove of garlic, peeled and finely minced

Pinch of salt and pepper
½ teaspoon of lemon juice
¼ cup of shredded cheddar cheese
¼ cup of shredded mozzarella cheese

Instructions:

1 Preheat the Innsky air fryer to 350 degrees.
In the oven, grill the tilapia with a little salt and lemon slices in foil on high heat for 20 minutes. Remove fish in foil from the oven, and break the fish meat apart into bite-sized pieces with a fork – it should be flaky and chunky when cooked. While the fish is cooling, combine the avocado, garlic, salt, pepper, and lemon juice in a small mixing bowl; mash lightly, but don't whip - keep the avocado slightly chunky. Spread the guacamole on one of the tortillas, then cover with the fish flakes, and then with the cheese. Top with the second tortilla. Place directly on hot surface of the air frying basket.

2 Set the Innsky air fryer timer for 6 minutes. After 6 minutes, when the air-fryer shuts off, flip the tortillas onto the other side with a spatula; the cheese should be melted enough that it won't fall apart. Reset air fryer to 350 degrees for another 6 minutes. After 6 minutes, when the air fryer shuts off, the tortillas should be browned and crisp, and the fish, guacamole and cheese will be hot and delicious inside. Remove with spatula and let sit on a serving plate to cool for a few minutes before slicing.

Crispy Paprika Fish Fillets

PREP: 5 MINUTES • COOK TIME: 15 MINUTES • TOTAL: 20 MINUTES • *SERVES: 4*

Ingredients

1/2 cup seasoned breadcrumbs
1 tablespoon balsamic vinegar
1/2 teaspoon seasoned salt
1 teaspoon paprika

1/2 teaspoon ground black pepper
1 teaspoon celery seed
2 fish fillets, halved
1 egg, beaten

Instructions:

1 Add the breadcrumbs, vinegar, salt, paprika, ground black pepper, and celery seeds to your food processor. Process for about 30 seconds.
Coat the fish fillets with the beaten egg; then, coat them with the breadcrumbs mixture.

2 Cook at 350 degrees F for about 15 minutes.

Parmesan Shrimp

PREP: 5 MINUTES • COOK TIME: 10 MINUTES • TOTAL: 15 MINUTES • *SERVES: 4*

Ingredients

2 tbsp. olive oil
1 tsp. onion powder
1 tsp. basil
½ tsp. oregano

1 tsp. pepper
2/3 C. grated parmesan cheese
4 minced garlic cloves
pounds of jumbo cooked shrimp (peeled/deveined)

Instructions:

1 Mix all seasonings together and gently toss shrimp with mixture.
2 Spray olive oil into the Innsky air fryer basket and add seasoned shrimp.
Cook 8-10 minutes at 350 degrees. Squeeze lemon juice over shrimp right before devouring

Quick Fried Catfish

PREP: 5 MINUTES • COOK TIME: 15 MINUTES • TOTAL: 20 MINUTES • *SERVES: 4*

Ingredients

3/4 cups Original Bisquick™ mix
1/2 cup yellow cornmeal
1 tablespoon seafood seasoning

4 catfish fillets (4 to 6 ounces each)
1/2 cup ranch dressing
Lemon wedges

Instructions:

1. In a shallow bowl mix together the Bisquick mix, cornmeal, and seafood seasoning. Pat the filets dry, then brush them with ranch dressing. Press the filets into the Bisquick mix on both sides until the filet is evenly coated.
2. Cook in your air fryer at 360 degrees for 15 minutes, flip the filets halfway through. Serve with a lemon garnish.

Per Serving: Calories: 372; Fat:16g; Protein:28g; Fiber:1.7g

Honey Glazed Salmon

PREP: 5 MINUTES • COOK TIME: 8 MINUTES • TOTAL: 13 MINUTES • *SERVES: 2*

Ingredients

1 tsp. water
3 tsp. rice wine vinegar
6 tbsp. low-sodium soy sauce

6 tbsp. raw honey
2 salmon fillets

Instructions:

1 Combine water, vinegar, honey, and soy sauce together. Pour half of this mixture into a bowl. Place salmon in one bowl of marinade and let chill 2 hours.
2 Ensure your Air fryer is preheated to 356 degrees and add salmon.
   Cook 8 minutes, flipping halfway through. Baste salmon with some of the remaining marinade mixture and cook another 5 minutes. To make a sauce to serve salmon with, pour remaining marinade mixture into a saucepan, heating till simmering. Let simmer 2 minutes. Serve drizzled over salmon!

Per Serving: Calories: 348; Fat:12g; Protein:20g; Sugar:3g

Fish Sandwiches

PREP: 10 MINUTES • COOK TIME: 20 MINUTES • TOTAL: 30 MINUTES • *SERVES: 4*

Ingredients

lbs White Fish Fillets
1/4 Cup Yellow Cornmeal
1 Tsp Greek Seasoning
Salt and Pepper to taste
2 ½ Cups Plain Flour
2 Tsps Baking Powder

2 Cups Beer
4 Hamburger Buns
Mayonnaise
Lettuce Leaves
1 Tomato, sliced
1 Egg

Instructions:

1 Cut the fish fillets into burger patty sized strips. Season with salt and pepper to desired taste.
   In a medium bowl, mix together the beer, egg, baking powder, plain flour, cornmeal, Greek seasoning and additional salt and pepper
   Heat the Air fryer to 340 degrees
   Place each seasoned fish strip into the batter, ensuring that it is well coated
2 Place battered fish into the Air fryer tray and cook in batches for 6 minutes or until crispy
   Compile the sandwich by topping each bun with mayonnaise, then a lettuce leaf, tomato slices, and finally the cooked fish strip

Crab Cakes

PREP: 5 MINUTES • COOK TIME: 10 MINUTES • TOTAL: 15 MINUTES • *SERVES: 4*

Ingredients

8 ounces jumbo lump crabmeat
1 tablespoon Old Bay Seasoning
⅓ cup bread crumbs
¼ cup diced red bell pepper
¼ cup diced green bell pepper

1 egg
¼ cup mayonnaise
Juice of ½ lemon
1 teaspoon flour
Cooking oil

Instructions:

1   In a large bowl, combine the crabmeat, Old Bay Seasoning, bread crumbs, red bell pepper, green bell pepper, egg, mayo, and lemon juice. Mix gently to combine. Form the mixture into 4 patties. Sprinkle ¼ teaspoon of flour on top of each patty.

2   Place the crab cakes in the air fryer. Spray them with cooking oil. Cook for 10 minutes. Serve.

Fish and Chips

PREP: 10 MINUTES • COOK TIME: 20 MINUTES • TOTAL: 30 MINUTES • *SERVES: 4*

Ingredients

4 (4-ounce) fish fillets
Pinch salt
Freshly ground black pepper
½ teaspoon dried thyme

1 egg white
¾ cup crushed potato chips
2 tablespoons olive oil, divided
3 russet potatoes, peeled and cut into strips

Instructions:

1   Pat the fish fillets dry and sprinkle with salt, pepper, and thyme. Set aside. In a shallow bowl, beat the egg white until foamy. In another bowl, combine the potato chips and 1 tablespoon of olive oil and mix until combined. Dip the fish fillets into the egg white, then into the crushed potato chip mixture to coat. Toss the fresh potato strips with the remaining 1 tablespoon olive oil.

2   Use your separator to divide the Innsky air fryer basket in half, then fry the chips and fish. The chips will take about 20 minutes; the fish will take about 10 to 12 minutes to cook.

Crispy Air Fried Sushi Roll

PREP: 10 MINUTES • COOK TIME: 5 MINUTES • TOTAL: 15 MINUTES • SERVES: 12

Ingredients

Kale Salad:
1 tbsp. sesame seeds
¾ tsp. soy sauce
¼ tsp. ginger
1/8 tsp. garlic powder
¾ tsp. toasted sesame oil
½ tsp. rice vinegar
1 ½ C. chopped kale
Sushi Rolls:

½ of a sliced avocado
3 sheets of sushi nori
1 batch cauliflower rice
Sriracha Mayo:
Sriracha sauce
¼ C. vegan mayo
Coating:
½ C. panko breadcrumbs

Instructions:

1   Combine all of kale salad ingredients together, tossing well. Set to the side. Lay out a sheet of nori and spread a handful of rice on. Then place 2-3 tbsp. of kale salad over rice, followed by avocado. Roll up sushi. To make mayo, whisk mayo ingredients together until smooth. Add breadcrumbs to a bowl.

2   Coat sushi rolls in crumbs till coated and add to the Air fryer. Cook rolls 10 minutes at 390 degrees, shaking gently at 5 minutes. Slice each roll into 6-8 pieces and enjoy.

Per Serving: Calories: 267; Fat:13g; Protein:6g; Sugar:3g

Sweet Recipes

Perfect Cinnamon Toast
PREP: 10 MINUTES • COOK TIME: 5 MINUTES • TOTAL: 15 MINUTES • *SERVES: 6*

Ingredients

2 tsp. pepper

1 ½ tsp. vanilla extract

1 ½ tsp. cinnamon

½ C. sweetener of choice

1 C. coconut oil

12 slices whole wheat bread

Instructions:

1   Melt coconut oil and mix with sweetener until dissolved. Mix in remaining ingredients minus bread till incorporated. Spread mixture onto bread, covering all area.
2   Place coated pieces of bread in your Air fryer.
Cook 5 minutes at 400 degrees. Remove and cut diagonally. Enjoy.

Per Serving: Calories: 124; Fat:2g; Protein:0g; Sugar:4g

Easy Baked Chocolate Mug Cake
PREP: 5 MINUTES • COOK TIME: 15 MINUTES • TOTAL: 20 MINUTES • *SERVES: 3*

Ingredients

½ cup cocoa powder

½ cup stevia powder

1 cup coconut cream

1 package cream cheese, room temperature

1 tablespoon vanilla extract

4 tablespoons butter

Instructions:

1   Preheat the Innsky air fryer for 5 minutes. In a mixing bowl, combine all ingredients. Use a hand mixer to mix everything until fluffy. Pour into greased mugs. Place the mugs in the fryer basket.
2   Bake for 15 minutes at 350°F.Place in the fridge to chill before serving.

Per Serving: Calories: 744; Fat:69.7g; Protein:13.9g; Sugar:4g

Easy Chocolate-Frosted Doughnuts
PREP: 5 MINUTES • COOK TIME: 5 MINUTES • TOTAL: 10 MINUTES • *SERVES: 6*

Ingredients

1 (16.3-ounce / 8-count) package refrigerated biscuit dough

¾ cup powdered sugar

¼ cup unsweetened cocoa powder

¼ cup milk

Instructions:

1   Spray the Innsky air fryer basket with olive oil.
Unroll the biscuit dough onto a cutting board and separate the biscuits.
Using a 1-inch biscuit cutter or cookie cutter, cut out the center of each biscuit.
Place the doughnuts into the air fryer. (You may have to cook your doughnuts in more than one batch.)
2   Set the temperature of your Innsky AF to 330°F. Set the timer and bake for 5 minutes.
Using tongs, remove the doughnuts from the air fryer and let them cool slightly before glazing.
Meanwhile, in a small mixing bowl, combine the powdered sugar, unsweetened cocoa powder, and milk and mix until smooth.
Dip your doughnuts into the glaze and use a knife to smooth the frosting evenly over the doughnut.
Let the glaze set before serving.

Per Serving (1 doughnut): Calories: 233; Fat: 8g; Saturated fat: 3g; Carbohydrate: 37g; Fiber: 2g; Sugar: 15g; Protein: 5g; Iron: 2mg; Sodium: 590mg

Air Fryer Homemade Pumpkin Fritters

PREP: 5 MINUTES • COOK TIME: 9 MINUTES • TOTAL: 14 MINUTES • *SERVES: 8 FRITTERS*

Ingredients

FOR THE FRITTERS

1 (16.3-ounce, 8-count) package refrigerated biscuit dough

½ cup chopped pecans

¼ cup pumpkin purée

¼ cup sugar

1 teaspoon pumpkin pie spice

2 tablespoons unsalted butter, melted

FOR THE GLAZE

1 cup powdered sugar

1 teaspoon pumpkin pie spice

1 tablespoon pumpkin purée

2 tablespoons milk (plus more to thin the glaze, if necessary)

Instructions:

1   TO MAKE THE FRITTERS. Spray the Innsky air fryer basket with olive oil or spray an air fryer–size baking sheet with olive oil or cooking spray. Turn the biscuit dough out onto a cutting board. Cut each biscuit into 8 pieces. Once you cut all the pieces, place them in a medium mixing bowl. Add the pecans, pumpkin, sugar, and pumpkin pie spice to the biscuit pieces and toss until well combined. Shape the dough into 8 even mounds. Drizzle butter over each of the fritters.

2   Place the fritters directly in the greased air fryer basket, or on the greased baking sheet set in the air fryer basket. Set the temperature of your Innsky AF to 330°F. Set the timer and bake for 7 minutes.

Check to see if the fritters are done. The dough should be cooked through and solid to the touch. If not, cook for 1 to 2 minutes more. Using tongs, gently remove the fritters from the air fryer. Let cool for about 10 minutes before you apply the glaze.

TO MAKE THE GLAZE

In a small mixing bowl, mix together the powdered sugar, pumpkin pie spice, pumpkin, and milk until smooth. If it seems more like icing, it is too thick. It should coat a spoon and be of a pourable consistency.

Drizzle the glaze over the fritters.

Per Serving (1 fritter): Calories: 341; Fat: 16g; Saturated fat: 5g; Carbohydrate: 47g; Fiber: 2g; Sugar: 26g; Protein: 5g; Iron: 2mg; Sodium: 608mg

Angel Food Cake

PREP: 5 MINUTES • COOK TIME: 30 MINUTES • TOTAL: 35 MINUTES • *SERVES: 12*

Ingredients

¼ cup butter, melted

1 cup powdered erythritol

1 teaspoon strawberry extract

12 egg whites

2 teaspoons cream of tartar

A pinch of salt

Instructions:

1   Preheat the Innsky air fryer for 5 minutes.

Mix the egg whites and cream of tartar. Use a hand mixer and whisk until white and fluffy.

Add the rest of the ingredients except for the butter and whisk for another minute.

Pour into a baking dish.

2   Place in the Innsky air fryer basket and cook for 30 minutes at 400°F or if a toothpick inserted in the middle comes out clean. Drizzle with melted butter once cooled.

Per Serving: Calories: 65; Fat:5g; Protein:3.1g; Fiber:1g

Air Fryer Homemade Chocolate Chip Cookies

PREP: 5 MINUTES • COOK TIME: 5 MINUTES • TOTAL: 10 MINUTES • *SERVES: 25 COOKIES*

Ingredients

1 cup (2 sticks) unsalted butter, at room temperature
1 cup granulated sugar
1 cup brown sugar
2 large eggs
½ teaspoon vanilla extract

1 teaspoon baking soda
½ teaspoon salt
3 cups all-purpose flour
2 cups chocolate chips

Instructions:

1 Spray an air fryer–size baking sheet with cooking spray

In a large bowl, cream the butter and both sugars. Mix in the eggs, vanilla, baking soda, salt, and flour until well combined. Fold in the chocolate chips. Use your hands and knead the dough together, so everything is well mixed. Using a cookie scoop or a tablespoon, drop heaping spoonfuls of dough onto the baking sheet about 1 inch apart Set the baking sheet into the air fryer. Set the temperature of your Innsky AF to 340°F. Set the timer and bake for 5 minutes. When the cookies are golden brown and cooked through, use silicone oven mitts to remove the baking sheet from the air fryer and serve. If you line your Innsky air fryer basket with air fryer parchment paper sprayed with cooking spray, you can cook multiple batches of cookies with very little cleanup.

Fried Peaches

PREP: 2 HOURS 10 MINUTES • COOK TIME: 15 MINUTES • TOTAL: 15 MINUTES • *SERVES: 4*

Ingredients

4 ripe peaches (1/2 a peach = 1 serving)
1 1/2 cups flour
Salt
2 egg yolks
3/4 cups cold water

1 1/2 tablespoons olive oil
2 tablespoons brandy
4 egg whites
Cinnamon/sugar mix

Instructions:

1 Mix flour, egg yolks, and salt in a mixing bowl. Slowly mix in water, then add brandy. Set the mixture aside for 2 hours and go do something for 1 hour 45 minutes.

Boil a large pot of water and cut and X at the bottom of each peach. While the water boils fill another large bowl with water and ice. Boil each peach for about a minute, then plunge it in the ice bath. Now the peels should basically fall off the peach. Beat the egg whites and mix into the batter mix. Dip each peach in the mix to coat.

2 Cook at 360 degrees for 10 Minutes.

Prepare a plate with cinnamon/sugar mix, roll peaches in mix and serve.

Per Serving: Calories: 306; Fat:3g; Protein:10g; Fiber:2.7g

Apple Dumplings

PREP: 10 MINUTES • COOK TIME: 25 MINUTES • TOTAL: 35 MINUTES • *SERVES: 4*

Ingredients

2 tbsp. melted coconut oil
2 puff pastry sheets
1 tbsp. brown sugar

2 tbsp. raisins
2 small apples of choice

Instructions:

1 Ensure your air fryer is preheated to 356 degrees.

Core and peel apples and mix with raisins and sugar.

Place a bit of apple mixture into puff pastry sheets and brush sides with melted coconut oil.

2 Place into the Air fryer. Cook 25 minutes, turning halfway through. Will be golden when done.

Air Fryer Stuffed Baked Apples
PREP: 5 MINUTES • COOK TIME: 20 MINUTES • TOTAL: 15 MINUTES • *SERVES: 4*

Ingredients

4 to 6 tablespoons chopped walnuts

4 to 6 tablespoons raisins

4 tablespoons (½ stick) unsalted butter, melted

1 teaspoon ground cinnamon

½ teaspoon ground nutmeg

4 apples, cored but with the bottoms left intact

Vanilla ice cream, for topping

Maple syrup, for topping

Instructions:

1 In a small mixing bowl, make the filling. Mix together the walnuts, raisins, melted butter, cinnamon, and nutmeg. Scoop a quarter of the filling into each apple. Place the apples in an air fryer–safe pan and set the pan in the air fryer basket.

2 Set the temperature of your Innsky AF to 350°F. Set the timer and bake for 20 minutes.
Serve with vanilla ice cream and a drizzle of maple syrup. Variation Tip: If you'd like to make baked apples with oatmeal filling, just add 1 cup of rolled oats and ¼ cup of brown sugar to the filling.

Per Serving: Calories: 382; Fat: 19g; Saturated fat: 9g; Carbohydrate: 57g; Fiber: 7g; Sugar: 44g; Protein: 4g; Iron: 2mg; Sodium: 100mg

Easy Air Fried Apple Hand Pies
PREP: 5 MINUTES • COOK TIME: 7 MINUTES • TOTAL: 12 MINUTES • *SERVES: 8 HAND PIES*

Ingredients

1 package prepared pie dough

½ cup apple pie filling

1 large egg white

1 tablespoon Wilton White Sparkling Sugar

Caramel sauce, for drizzling

Instructions:

1 Spray the Innsky air fryer basket with olive oil. Lightly flour a clean work surface. Lay out the dough on the work surface. Using a 2-inch biscuit cutter, cut out 8 circles from the dough. Gather up the scraps of dough, form them into a ball, and reroll them. Using the biscuit cutter, cut out the remaining dough. Add about 1 tablespoon of apple pie filling to the center of each circle. Fold over the dough and use a fork to seal the edges. Brush the egg white over the top, then sprinkle with sparkling sugar. Place the hand pies in the greased air fryer basket. They should be spaced so that they do not touch one another.

2 Set the temperature of your Innsky AF to 350°F. Set the timer and bake for 5 minutes. When they are done, the crust should be golden brown. If they are not done, bake for another 2 minutes. Drizzle with caramel sauce, if desired.

Per Serving (1 pie): Calories: 120; Fat: 5g; Saturated fat: 1g; Carbohydrate: 17g; Fiber: 0g; Sugar: 3g; Protein: 1g; Iron: 0mg; Sodium: 144mg

Apple Pie in Air Fryer
PREP: 5 MINUTES • COOK TIME: 35 MINUTES • TOTAL: 40 MINUTES • *SERVES: 4*

Ingredients

½ teaspoon vanilla extract
1 beaten egg
1 large apple, chopped
1 Pillsbury Refrigerator pie crust
1 tablespoon butter

1 tablespoon ground cinnamon
1 tablespoon raw sugar
2 tablespoon sugar
2 teaspoons lemon juice
Baking spray

Instructions:

1 Lightly grease baking pan of air fryer with cooking spray. Spread pie crust on bottom of pan up to the sides. In a bowl, mix vanilla, sugar, cinnamon, lemon juice, and apples. Pour on top of pie crust. Top apples with butter slices. Cover apples with the other pie crust. Pierce with knife the tops of pie. Spread beaten egg on top of crust and sprinkle sugar. Cover with foil.

2 For 25 minutes, cook on 390°F. Remove foil cook for 10 minutes at 330oF until tops are browned. Serve and enjoy.

Easy Air Fryer Blueberry Pie
PREP: 5 MINUTES PLUS 30 MINUTES TO THAW • COOK TIME: 18 MINUTES • TOTAL: 53 MINUTES •*SERVES: 4-6*

Ingredients

2 frozen pie crusts
2 (21-ounce) jars blueberry pie filling
1 teaspoon milk

1 teaspoon sugar

Instructions:

1 Remove the pie crusts from the freezer and let them thaw for 30 minutes on the countertop. Place one crust into the bottom of a 6-inch pie pan. Pour the pie filling into the bottom crust, then cover it with the other crust, being careful to press the bottom and top crusts together around the edge to form a seal. Trim off any excess pie dough. Cut venting holes in the top crust with a knife or a small decoratively shaped cookie cutter. Brush the top crust with milk, then sprinkle the sugar over it. Place the pie in the air fryer basket.

2 Set the temperature of your Innsky AF to 310°F. Set the timer and bake for 15 minutes.
Check the pie after 15 minutes. If it needs additional time, reset the timer and bake for an additional 3 minutes. Using silicone oven mitts, remove the pie from the air fryer and let cool for 15 minutes before serving.

Air Fryer Chocolate Cake
PREP: 5 MINUTES • COOK TIME: 35 MINUTES • TOTAL: 40 MINUTES • *SERVES: 8-10*

Ingredients

½ C. hot water
1 tsp. vanilla
¼ C. olive oil
½ C. almond milk
1 egg
½ tsp. salt

¾ tsp. baking soda
¾ tsp. baking powder
½ C. unsweetened cocoa powder
2 C. almond flour
1 C. brown sugar

Instructions:

1 Preheat your air fryer to 356 degrees.
Stir all dry ingredients together. Then stir in wet ingredients. Add hot water last.
The batter will be thin, no worries.

2 Pour cake batter into a pan that fits into the fryer. Cover with foil and poke holes into the foil. Bake 35 minutes. Discard foil and then bake another 10 minutes.

Per Serving: Calories: 378; Fat:9g; Protein:4g; Sugar:5g

Raspberry Cream Rol-Ups
PREP: 10 MINUTES • COOK TIME: 25 MINUTES • TOTAL: 35 MINUTES • *SERVES: 4*

Ingredients
1 cup of fresh raspberries, rinsed and patted dry
½ cup of cream cheese, softened to room temperature
¼ cup of brown sugar
¼ cup of sweetened condensed milk

1 egg
1 teaspoon of corn starch
6 spring roll wrappers
¼ cup of water

Instructions:
1   Cover the basket of the Air fryer with a lining of tin foil, leaving the edges uncovered to allow air to circulate through the basket. Preheat the Innsky air fryer to 350 degrees. In a mixing bowl, combine the cream cheese, brown sugar, condensed milk, cornstarch, and egg. Beat or whip thoroughly, until all ingredients are completely mixed and fluffy, thick and stiff. Spoon even amounts of the creamy filling into each spring roll wrapper, then top each dollop of filling with several raspberries. Roll up the wraps around the creamy raspberry filling, and seal the seams with a few dabs of water. Place each roll on the foil-lined Air fryer basket, seams facing down.

2   Set the Innsky air fryer timer to 10 minutes. During cooking, shake the handle of the fryer basket to ensure a nice even surface crisp. After 10 minutes, when the Air fryer shuts off, the spring rolls should be golden brown and perfect on the outside, while the raspberries and cream filling will have cooked together in a glorious fusion. Remove with tongs and serve hot or cold.

Air Fryer Banana Cake
PREP: 5 MINUTES • COOK TIME: 30 MINUTES • TOTAL: 35 MINUTES • *SERVES: 4*

Ingredients
⅓ cup brown sugar
4 tablespoons (½ stick) unsalted butter, at room temperature
1 ripe banana, mashed
1 large egg

2 tablespoons granulated sugar
1 cup all-purpose flour
1 teaspoon ground cinnamon
1 teaspoon vanilla extract
½ teaspoon ground nutmeg

Instructions:
1   Spray a 6-inch Bundt pan with cooking spray.
In a medium mixing bowl, cream the brown sugar and butter until pale and fluffy.
Mix in the banana and egg. Add the granulated sugar, flour, ground cinnamon, vanilla, and nutmeg and mix well. Spoon the batter into the prepared pan. Place the pan in the air fryer basket.

2   Set the temperature of your Innsky AF to 320°F. Set the timer and bake for 15 minutes.
Do a toothpick test. If the toothpick comes out clean, the cake is done. It there is batter on the toothpick, cook and check again in 5-minute intervals until the cake is done. It will likely take about 30 minutes total baking time to fully cook. Using silicone oven mitts, remove the Bundt pan from the air fryer. Set the pan on a wire cooling rack and let cool for about 10 minutes. Place a plate upside-down (like a lid) over the top of the Bundt pan. Carefully flip the plate and the pan over, and set the plate on the counter. Lift the Bundt pan off the cake. Frost as desired.

Per Serving: Calories: 334; Fat: 13g; Saturated fat: 8g; Carbohydrate: 49g; Fiber: 2g; Sugar: 22g; Protein: 5g; Iron: 2mg; Sodium: 104mg

Banana-Choco Brownies
PREP: 5 MINUTES • COOK TIME: 30 MINUTES • TOTAL: 35 MINUTES • *SERVES: 12*

Ingredients

2 cups almond flour
2 teaspoons baking powder
½ teaspoon baking powder
½ teaspoon baking soda
½ teaspoon salt
1 over-ripe banana

3 large eggs
½ teaspoon stevia powder
¼ cup coconut oil
1 tablespoon vinegar
1/3 cup almond flour
1/3 cup cocoa powder

Instructions:

1 Preheat the Innsky air fryer for 5 minutes. Combine all ingredients in a food processor and pulse until well-combined. Pour into a baking dish that will fit in the air fryer.

2 Place in the Innsky air fryer basket and cook for 30 minutes at 350°F or if a toothpick inserted in the middle comes out clean.

Per Serving: Calories: 75; Fat:6.5g; Protein:1.7g; Sugar:2g

Easy Air Fried Old-Fashioned Cherry Cobbler
PREP: 5 MINUTES • COOK TIME: 35 MINUTES • TOTAL: 15 MINUTES • *SERVES: 4*

Ingredients

1 cup all-purpose flour
1 cup sugar
2 tablespoons baking powder

¾ cup milk
8 tablespoons (1 stick) unsalted butter
1 (21-ounce) can cherry pie filling

Instructions:

1 In a small mixing bowl, mix together the flour, sugar, and baking powder. Add the milk and mix until well blended. Melt the butter in a small microwave-safe bowl in the microwave, about 45 seconds. Pour the butter into the bottom of an 8-by-8-inch pan, then pour in the batter and spread it in an even layer. Pour the pie filing over the batter. Do not mix; the batter will bubble up through the filling durin cooking.

2 Set the temperature of your Innsky AF to 320°F. Set the timer and bake for 20 minutes.
Check the cobbler. When the cobbler is done the batter will be golden brown and cooked through. If not done, bake and recheck for doneness in 5-minute intervals. Overall cooking time will likely be between 30 and 35 minutes. Remove from the air fryer and let cool slightly before serving.

Per Serving: Calories: 706; Fat: 24g; Saturated fat: 15g; Carbohydrate: 121g; Fiber: 2g; Sugar: 52g; Protein: 6g; Iron: 2mg; Sodium: 219mg

Chocolate Donuts
PREP: 5 MINUTES • COOK TIME: 20 MINUTES • TOTAL: 25 MINUTES • *SERVES: 8-10*

Ingredients

(8-ounce) can jumbo biscuits
Cooking oil

Chocolate sauce, such as Hershey's

Instructions:

1 Separate the biscuit dough into 8 biscuits and place them on a flat work surface. Use a small circle cookie cutter or a biscuit cutter to cut a hole in the center of each biscuit. You can also cut the holes using a knife.
Spray the Innsky air fryer basket with cooking oil.

2 Place 4 donuts in the air fryer. Do not stack. Spray with cooking oil. Cook for 4 minutes.
Open the air fryer and flip the donuts. Cook for an additional 4 minutes.
Remove the cooked donuts from the air fryer, then repeat steps 3 and 4 for the remaining 4 donuts.
Drizzle chocolate sauce over the donuts and enjoy while warm.

Per Serving: Calories: 181; Fat:98g; Protein:3g; Fiber:1g

Homemade Air Fried Fudge Brownies
PREP: 5 MINUTES • COOK TIME: 20 MINUTES • TOTAL: 25 MINUTES • *SERVES: 6*

Ingredients

8 tablespoons (1 stick) unsalted butter, melted
1 cup sugar
1 teaspoon vanilla extract
2 large eggs

½ cup all-purpose flour
½ cup cocoa powder
1 teaspoon baking powder

Instructions:

1 Spray a 6-inch air fryer–safe baking pan with cooking spray or grease the pan with butter. In a medium mixing bowl, mix together the butter and sugar, then add the vanilla and eggs and beat until well combined. Add the flour, cocoa powder, and baking powder and mix until smooth. Pour the batter into the prepared pan.

2 Set the temperature of your Innsky AF to 350°F. Set the timer and bake for 20 minutes. Once the center is set, use silicon oven mitts to remove the pan from the air fryer. Let cool slightly before serving.

Chocolate Bundt Cake
PREP: 5 MINUTES • COOK TIME: 30 MINUTES • TOTAL: 35 MINUTES • *SERVES: 4*

Ingredients

1¾ cups all-purpose flour
2 cups sugar
¾ cup unsweetened cocoa powder
1 teaspoon baking soda
1 teaspoon baking powder
½ cup vegetable oil

1 teaspoon salt
2 teaspoons vanilla extract
2 large eggs
1 cup milk
1 cup hot water

Instructions:

1 Spray a 6-inch Bundt pan with cooking spray.
In a large mixing bowl, combine the flour, sugar, cocoa powder, baking soda, baking powder, oil, salt, vanilla, eggs, milk, and hot water. Pour the cake batter into the prepared pan and set the pan in the air fryer basket.

2 Set the temperature of your Innsky AF to 330°F. Set the timer and bake for 20 minutes. Do a toothpick test. If the toothpick comes out clean, the cake is done. It there is batter on the toothpick, cook and check again in 5-minute intervals until the cake is done. It will likely take about 30 minutes total baking time to fully cook. Using silicone oven mitts, remove the Bundt pan from the air fryer. Set the pan on a wire cooling rack and let cool for about 10 minutes. Place a plate upside down over the top of the Bundt pan. Carefully flip the plate and the pan over, and set the plate on the counter. Lift the Bundt pan off the cake.

Per Serving: Calories: 924; Fat: 34g; Saturated fat: 6g; Carbohydrate: 155g; Fiber: 6g; Sugar: 104g; Protein: 14g; Iron: 6mg; Sodium: 965mg

Easy Air Fryer Donuts
PREP: 5 MINUTES • COOK TIME: 5 MINUTES • TOTAL: 10 MINUTES • *SERVES: 8*

Ingredients

Pinch of allspice
4 tbsp. dark brown sugar
½ - 1 tsp. cinnamon

1/3 C. granulated sweetener
3 tbsp. melted coconut oil
1 can of biscuits

Instructions:

1 Mix allspice, sugar, sweetener, and cinnamon together.
Take out biscuits from can and with a circle cookie cutter, cut holes from centers and place into air fryer.

2 Cook 5 minutes at 350 degrees. As batches are cooked, use a brush to coat with melted coconut oil and dip each into sugar mixture.
Serve warm.

Little French Fudge Cakes

PREP: 10 MINUTES • COOK TIME: 25 MINUTES • TOTAL: 35 MINUTES • *SERVES: 12 CAKES*

Ingredients

3 cups blanched almond flour

¾ cup unsweetened cocoa powder

1 teaspoon baking soda

½ teaspoon fine sea salt

6 large eggs

1 cup Swerve confectioners'-style sweetener

1½ cups canned pumpkin puree

3 tablespoons brewed decaf espresso or other strong brewed decaf coffee

3 tablespoons unsalted butter, melted but not hot

1 teaspoon vanilla extract

CREAM CHEESE FROSTING:

½ cup Swerve confectioners'-style sweetener

½ cup (1 stick) unsalted butter, melted

4 ounces cream cheese (½ cup) softened

3 tablespoons unsweetened, unflavored almond milk or heavy cream

CHOCOLATE DRIZZLE:

3 tablespoons unsalted butter

2 tablespoons Swerve confectioners'-style sweetener or liquid sweetener

2 tablespoons unsweetened cocoa powder

¼ cup unsweetened, unflavored almond milk

½ cup chopped walnuts or pecans, for garnish (optional)

Instructions:

1   Preheat the Innsky air fryer to 350°F. Spray 2 mini Bundt pans with coconut oil.

In a medium-sized bowl, whisk together the flour, cocoa powder, baking soda, and salt until blended.

In a large bowl, beat the eggs and sweetener with a hand mixer for 2 to 3 minutes, until light and fluffy. Add the pumpkin puree, espresso, melted butter, and vanilla and stir to combine.

Add the wet ingredients to the dry ingredients and stir until just combined.

Pour the batter into the prepared pans, filling each well two-thirds full.

2   Cook in the air fryer for 20 to 25 minutes, until a toothpick inserted into the center of a cake comes out clean. Allow the cakes to cool completely in the pans before removing them. Make the frosting: In a large bowl, mix the sweetener, melted butter, and cream cheese until well combined. Add the almond milk and stir well to combine. Make the chocolate drizzle: In a small bowl, stir together the melted butter, sweetener, and cocoa powder until well combined. Add the almond milk while stirring to thin the mixture. After the cakes have cooled, dip the tops of the cakes into the frosting, then use a spoon to drizzle the chocolate over each frosted cake. If desired, garnish the cakes with chopped nuts.

Store leftovers in an airtight container in the refrigerator for up to 4 days or in the freezer for up to a month.

Chocolate Soufflé for Two

PREP: 5 MINUTES • COOK TIME: 14 MINUTES • TOTAL: 19 MINUTES • *SERVES: 2*

Ingredients

2 tbsp. almond flour

½ tsp. vanilla

3 tbsp. sweetener

2 separated eggs

¼ C. melted coconut oil

3 ounces of semi-sweet chocolate, chopped

Instructions:

1   Brush coconut oil and sweetener onto ramekins.

Melt coconut oil and chocolate together.

Beat egg yolks well, adding vanilla and sweetener. Stir in flour and ensure there are no lumps. Preheat fryer to 330 degrees. Whisk egg whites till they reach peak state and fold them into chocolate mixture. Pour batter into ramekins and place into the fryer.

2   Cook 14 minutes. Serve with powdered sugar dusted on top.

Per Serving: Calories: 238; Fat:6g; Protein:1g; Sugar:4g

Fried Bananas with Chocolate Sauce
PREP: 10 MINUTES • COOK TIME: 10 MINUTES • TOTAL: 20 MINUTES • *SERVES: 2*

Ingredients
1 large egg
¼ cup cornstarch
¼ cup plain bread crumbs

3 bananas, halved crosswise
Cooking oil
Chocolate sauce

Instructions:
1   In a small bowl, beat the egg. In another bowl, place the cornstarch. Place the bread crumbs in a third bowl. Dip the bananas in the cornstarch, then the egg, and then the bread crumbs.
Spray the Innsky air fryer basket with cooking oil. Place the bananas in the basket and spray them with cooking oil.
2   Cook for 5 minutes. Open the air fryer and flip the bananas. Cook for an additional 2 minutes. Transfer the bananas to plates.
Drizzle the chocolate sauce over the bananas, and serve.
You can make your own chocolate sauce using 2 tablespoons milk and ¼ cup chocolate chips. Heat a saucepan over medium-high heat. Add the milk and stir for 1 to 2 minutes. Add the chocolate chips. Stir for 2 minutes, or until the chocolate has melted.

Per Serving: Calories: 203; Fat:6g; Protein:3g; Fiber:3g

Flourless Cream-Filled Mini Cakes
PREP: 10 MINUTES • COOK TIME: 10 MINUTES • TOTAL: 20 MINUTES • *SERVES: 8*

Ingredients
Cake:
½ cup (1 stick) unsalted butter
4 ounces unsweetened chocolate, chopped
¾ cup Swerve confectioners'-style sweetener or equivalent amount of powdered sweetener
3 large eggs

Filling:
1 (8-ounce) package cream cheese softened
¼ cup Swerve confectioners'-style sweetener
Whipped cream
Raspberries

Instructions:
1   Preheat the Innsky air fryer to 375°F. Grease eight 4-ounce ramekins. Make the cake batter: Heat the butter and chocolate in a saucepan over low heat, stirring often, until the chocolate is completely melted. Remove from the heat.
Add the sweetener and eggs and use a hand mixer on low to combine well. Set aside.
Make the cream filling: In a medium-sized bowl, mix together the cream cheese and sweetener until well combined. Taste and add more sweetener if desired. Divide the chocolate mixture among the greased ramekins, filling each one halfway. Place 1 tablespoon of the filling on top of the chocolate mixture in each ramekin.
2   Place the ramekins in the air fryer and cook for 10 minutes, or until the outside is set and the inside is soft and warm. Allow to cool completely, then top with whipped cream, if desired, and garnish with raspberries, if desired.

Chocolaty Banana Muffins

PREP: 5 MINUTES • COOK TIME: 25 MINUTES • TOTAL: 35 MINUTES • *SERVES: 12*

Ingredients

¾ cup whole wheat flour
¾ cup plain flour
¼ cup cocoa powder
¼ teaspoon baking powder
1 teaspoon baking soda
¼ teaspoon salt

2 large bananas, peeled and mashed
1 cup sugar
1/3 cup canola oil
1 egg
½ teaspoon vanilla essence
1 cup mini chocolate chips

Instructions:

1   In a large bowl, mix together flour, cocoa powder, baking powder, baking soda and salt. In another bowl, add bananas, sugar, oil, egg and vanilla extract and beat till well combined. Slowly, add flour mixture in egg mixture and mix till just combined. Fold in chocolate chips. Preheat the Innsky air fryer to 345 degrees F. Grease 12 muffin molds.

2   Transfer the mixture into prepared muffin molds evenly and cook for about 20-25 minutes or till a toothpick inserted in the center comes out clean. Remove the muffin molds from Air fryer and keep on wire rack to cool for about 10 minutes. Carefully turn on a wire rack to cool completely before serving.

Halle Berries-and-Cream Cobbler

PREP: 10 MINUTES • COOK TIME: 25 MINUTES • TOTAL: 35 MINUTES • *SERVES: 4*

Ingredients

12 ounces cream cheese (1½ cups), softened
1 large egg
¾ cup Swerve confectioners'-style sweetener
½ teaspoon vanilla extract
¼ teaspoon fine sea salt
1 cup sliced fresh raspberries or strawberries
BISCUITS:
3 large egg whites
¾ cup blanched almond flour
1 teaspoon baking powder

2½ tablespoons very cold unsalted butter, cut into pieces
¼ teaspoon fine sea salt
FROSTING:
2 ounces cream cheese (¼ cup), softened
1 Tablespoon Swerve confectioners'-style sweetener or liquid sweetener
1 tablespoon unsweetened, unflavored almond milk or heavy cream
Fresh raspberries or strawberries, for garnish

Instructions:

1   Preheat the Innsky air fryer to 400°F. Grease a 7-inch pie pan.
In a large mixing bowl, use a hand mixer to combine the cream cheese, egg, and sweetener until smooth. Stir in the vanilla and salt. Gently fold in the raspberries with a rubber spatula. Pour the mixture into the prepared pan and set aside. Make the biscuits: Place the egg whites in a medium-sized mixing bowl or the bowl of a stand mixer. Using a hand mixer or stand mixer, whip the egg whites until very fluffy and stiff. In a separate medium-sized bowl, combine the almond flour and baking powder. Cut in the butter and add the salt, stirring gently to keep the butter pieces intact. Gently fold the almond flour mixture into the egg whites. Use a large spoon or ice cream scooper to scoop out the dough and form it into a 2-inch-wide biscuit, making sure the butter stays in separate clumps. Place the biscuit on top of the raspberry mixture in the pan. Repeat with remaining dough to make 4 biscuits.

2   Place the pan in the air fryer and cook for 5 minutes, then lower the temperature to 325°F and bake for another 17 to 20 minutes, until the biscuits are golden brown. While the cobbler cooks, make the frosting: Place the cream cheese in a small bowl and stir to break it up. Add the sweetener and stir. Add the almond milk and stir until well combined. If you prefer a thinner frosting, add more almond milk. Remove the cobbler from the air fryer and allow to cool slightly, then drizzle with the frosting. Garnish with fresh raspberries.
Store leftovers in an airtight container in the refrigerator for up to 3 days. Reheat the cobbler in a preheated 350°F air fryer for 3 minutes, or until warmed through.

Apple Hand Pies

PREP: 5 MINUTES • COOK TIME: 8 MINUTES • TOTAL: 13 MINUTES • *SERVES: 6*

Ingredients
15-ounces no-sugar-added apple pie filling
1 store-bought crust

Instructions:
1   Lay out pie crust and slice into equal-sized squares.
Place 2 tbsp. filling into each square and seal crust with a fork.
2   Place into the Air fryer. Cook 8 minutes at 390 degrees until golden in color.

Chocolate Meringue Cookies

**PREP: 10 MINUTES PLUS 20 MINUTES TO REST • COOK TIME: 60 MINUTES • TOTAL: 1 HOUR 30 MINUTES
•*SERVES: 16 COOKIES***

Ingredients
3 large egg whites
¼ teaspoon cream of tartar

¼ cup Swerve confectioners'-style sweetener
2 tablespoons unsweetened cocoa powder

Instructions:

1. Preheat the Innsky air fryer to 225°F. Line a 7-inch pie pan or a dish that will fit in your air fryer with parchment paper. In a small bowl, use a hand mixer to beat the egg whites and cream of tartar until soft peaks form. With the mixer on low, slowly sprinkle in the sweetener and mix until it's completely incorporated. Continue to beat with the mixer until stiff peaks form. Add the cocoa powder and gently fold until it's completely incorporated. Spoon the mixture into a piping bag with a ¾-inch tip. (If you don't have a piping bag, snip the corner of a large resealable plastic bag to form a ¾-inch hole.) Pipe sixteen 1-inch meringue cookies onto the lined pie pan, spacing them about ¼ inch apart.
2. Place the pan in the air fryer and cook for 1 hour, until the cookies are crispy on the outside, then turn off the air fryer and let the cookies stand in the air fryer for another 20 minutes before removing and serving.

Lemon Poppy Seed Macaroons

PREP: 10 MINUTES • COOK TIME: 14 MINUTES • TOTAL: 24 MINUTES • *SERVES: 12 COOKIES*

Ingredients
2 large egg whites, room temperature
⅓ cup Swerve confectioners'-style sweetener
2 tablespoons grated lemon zest, plus more for
garnish if desired
2 teaspoons poppy seeds
1 teaspoon lemon extract

¼ teaspoon fine sea salt
2 cups unsweetened shredded coconut
LEMON ICING:
¼ cup Swerve confectioners'-style sweetener
1 tablespoon lemon juice

Instructions:
1   Preheat the Innsky air fryer to 325°F. Line a 7-inch pie pan or a casserole dish that will fit inside your air fryer with parchment paper. Place the egg whites in a medium-sized bowl and use a hand mixer on high to beat the whites until stiff peaks form. Add the sweetener, lemon zest, poppy seeds, lemon extract, and salt. Mix on low until combined. Gently fold in the coconut with a rubber spatula. Use a 1-inch cookie scoop to place the cookies on the parchment, spacing them about ¼ inch apart.
2   Place the pan in the air fryer and cook for 12 to 14 minutes, until the cookies are golden and a toothpick inserted into the center comes out clean. While the cookies bake, make the lemon icing: Place the sweetener in a small bowl. Add the lemon juice and stir well. If the icing is too thin, add a little more sweetener. If the icing is too thick, add a little more lemon juice. Remove the cookies from the air fryer and allow to cool for about 10 minutes, then drizzle with the icing. Garnish with lemon zest, if desired.

Per cookie: Calories 71; Fat 7g; Protein 1g; Total carbs 3g; Fiber 2g

Blueberry Lemon Muffins

PREP: 5 MINUTES • COOK TIME: 10 MINUTES • TOTAL: 15 MINUTES • *SERVES: 12*

Ingredients

1 tsp. vanilla
Juice and zest of 1 lemon
2 eggs
1 C. blueberries

½ C. cream
¼ C. avocado oil
½ C. monk fruit
2 ½ C. almond flour

Instructions:

1   Mix monk fruit and flour together.
    In another bowl, mix vanilla, egg, lemon juice, and cream together. Add mixtures together and blend well. Spoon batter into cupcake holders.
2   Place in air fryer. Bake 10 minutes at 320 degrees, checking at 6 minutes to ensure you don't overbake them.

Per Serving: Calories: 317; Fat:11g; Protein:3g; Sugar:5g

Lemon Curd Pavlova

PREP:10 MINUTES PLUS 20 MINUTES TO REST•COOK TIME:60 MINUTES•TOTAL: 1 HOUR 30 MINUTES•*SERVES: 4*

Ingredients

SHELL:
3 large egg whites
¼ teaspoon cream of tartar
¾ cup Swerve confectioners'-style sweetener
1 teaspoon grated lemon zest
1 teaspoon lemon extract
LEMON CURD:
1 cup Swerve confectioners'-style sweetener or powdered sweetener

½ cup lemon juice
4 large eggs
½ cup coconut oil
FOR GARNISH (OPTIONAL):
Blueberries
Swerve confectioners'-style sweetener or equivalent amount of powdered sweetener

Instructions:

1   Preheat the Innsky air fryer to 275°F. Thoroughly grease a 7-inch pie pan with butter or coconut oil.
    Make the shell: In a small bowl, use a hand mixer to beat the egg whites and cream of tartar until soft peaks form. With the mixer on low, slowly sprinkle in the sweetener and mix until it's completely incorporated.
    Add the lemon zest and lemon extract and continue to beat with the hand mixer until stiff peaks form.
    Spoon the mixture into the greased pie pan, then smooth it across the bottom, up the sides, and onto the rim to form a shell.
2   Cook for 1 hour, then turn off the air fryer and let the shell stand in the air fryer for 20 minutes. (The shell can be made up to 3 days ahead and stored in an airtight container in the refrigerator, if desired.)
    While the shell bakes, make the lemon curd: In a medium-sized heavy-bottomed saucepan, whisk together the sweetener, lemon juice, and eggs. Add the coconut oil and place the pan on the stovetop over medium heat. Once the oil is melted, whisk constantly until the mixture thickens and thickly coats the back of a spoon, about 10 minutes. Do not allow the mixture to come to a boil.
    Pour the lemon curd mixture through a fine-mesh strainer into a medium-sized bowl. Place the bowl inside a larger bowl filled with ice water and whisk occasionally until the curd is completely cool, about 15 minutes.
    Place the lemon curd on top of the shell and garnish with blueberries and powdered sweetener, if desired. Store leftovers in the refrigerator for up to 4 days.

Per serving: Calories 332; Fat 33g; Protein 9g; Total carbs 4g; Fiber 1g

Sweet Cream Cheese Wontons
PREP: 5 MINUTES • COOK TIME: 5 MINUTES • TOTAL: 10 MINUTES • *SERVES: 16*

Ingredients

1 egg mixed with a bit of water
Wonton wrappers
½ C. powdered erythritol

8 ounces softened cream cheese
Olive oil

Instructions:

1   Mix sweetener and cream cheese together.
    Lay out 4 wontons at a time and cover with a dish towel to prevent drying out.
    Place ½ of a teaspoon of cream cheese mixture into each wrapper. Dip finger into egg/water mixture and fold diagonally to form a triangle. Seal edges well. Repeat with remaining ingredients.
2   Place filled wontons into the Air fryer and cook 5 minutes at 400 degrees, shaking halfway through cooking.

Per Serving: Calories: 303; Fat:3g; Protein:0.5g; Sugar:4g

Air Fryer Cinnamon Rolls
PREP: 15 MINUTES • COOK TIME: 5 MINUTES • TOTAL: 15 MINUTES • *SERVES: 8*

Ingredients

1 ½ tbsp. cinnamon
¾ C. brown sugar
¼ C. melted coconut oil
1 pound frozen bread dough, thawed

Glaze:

½ tsp. vanilla
1 ¼ C. powdered erythritol
2 tbsp. softened ghee
3 ounces softened cream cheese

Instructions:

1   Lay out bread dough and roll out into a rectangle. Brush melted ghee over dough and leave a 1-inch border along edges. Mix cinnamon and sweetener together and then sprinkle over dough. Roll dough tightly and slice into 8 pieces. Let sit 1-2 hours to rise. To make the glaze, simply mix ingredients together till smooth.
2   Once rolls rise, place into air fryer and cook 5 minutes at 350 degrees.
    Serve rolls drizzled in cream cheese glaze. Enjoy.

Per Serving: Calories: 390; Fat:8g; Protein:1g; Sugar:7g

Blueberry–Cream Cheese Bread Pudding
PREP: 15 MINUTES • COOK TIME: 70 MINUTES • TOTAL: 1 HOUR 35 MINUTES • *SERVES: 6*

Ingredients

1 cup light cream or half-and-half
4 large eggs
⅓ cup plus 3 tablespoons granulated sugar
1 teaspoon pure lemon extract

4 cups cubed croissants (4 to 5 croissants)
1 cup blueberries
4 ounces cream cheese, cut into small cubes

Instructions:

1   In a large bowl, combine the cream, eggs, the ⅓ cup sugar, and the extract. Whisk until well combined. Add the cubed croissants, blueberries, and cream cheese. Toss gently until everything is thoroughly combined; set aside.
    Place a 3-cup Bundt pan in the air fryer basket. Preheat the Innsky air fryer to 400°F.
    Sprinkle the remaining 3 tablespoons sugar in the bottom of the hot pan.
2   Set the Innsky air fryer to 400°F for 10 minutes, or until the sugar caramelizes. Tip the pan to spread the caramel evenly across the bottom of the pan. Remove the pan from the air fryer and pour in the bread mixture, distributing it evenly across the pan. Place the pan in the air fryer basket. Set the Innsky air fryer to 350°F for 60 minutes, or until the custard is set in the middle. Let stand for 10 minutes before unmolding onto a serving plate.

Apple Dutch Baby

PREP: 20 MINUTES PLUS 30 MINUTES TO REST• COOK TIME: 16 MINUTES • TOTAL: 1HOR 5 MINUTES •*SERVES: 2-3*

Ingredients

For the Batter
2 large eggs
¼ cup all-purpose flour
¼ teaspoon baking powder
1½ teaspoons granulated sugar
Pinch kosher salt
½ cup whole milk
1 tablespoon butter, melted
½ teaspoon pure vanilla extract

¼ teaspoon ground nutmeg
For the Apples
2 tablespoon butter
4 tablespoons granulated sugar
¼ teaspoon ground cinnamon
¼ teaspoon ground nutmeg
1 small tart apple (such as Granny Smith), peeled, cored, and sliced
Vanilla ice cream (optional), for serving

Instructions:

1   For the batter: In a medium bowl, combine the eggs, flour, baking powder, sugar, and salt. Whisk lightly. While whisking continuously, slowly pour in the milk. Whisk in the melted butter, vanilla, and nutmeg. Let the batter stand for 30 minutes. For the apples: Place the butter in a 6 × 3-inch round heatproof pan. Place the pan in the air fryer basket. Set the Innsky air fryer to 400°F for 2 minutes. In a small bowl, combine 2 tablespoons of the sugar with the cinnamon and nutmeg and stir until well combined. When the pan is hot and the butter is melted, brush some butter up the sides of the pan. Sprinkle the spiced sugar mixture over the butter. Arrange the apple slices in the pan in a single layer and sprinkle the remaining 2 tablespoons sugar over the apples.

2   Set the Innsky air fryer to 400°F to 2 minutes, or until the mixture bubbles.
Gently pour the batter over the apples. Set the Innsky air fryer to 350°F for 12 minutes, or until the pancake is golden brown around the edges, the center is cooked through, and a toothpick emerges clean. Serve immediately with ice cream, if desired.

French Toast Bites

PREP: 5 MINUTES • COOK TIME: 15 MINUTES • TOTAL: 20 MINUTES • *SERVES: 8*

Ingredients

Almond milk
Cinnamon
Sweetener

3 eggs
4 pieces wheat bread

Instructions:

1   Preheat air fryer to 360 degrees. Whisk eggs and thin out with almond milk. Mix 1/3 cup of sweetener with lots of cinnamon. Tear bread in half, ball up pieces and press together to form a ball. Soak bread balls in egg and then roll into cinnamon sugar, making sure to thoroughly coat.

2   Place coated bread balls into the Air fryer and bake 15 minutes.

Coconutty Lemon Bars

PREP: 5 MINUTES • COOK TIME: 25 MINUTES • TOTAL: 30 MINUTES • *SERVES: 12*

Ingredients

¼ cup cashew
¼ cup fresh lemon juice, freshly squeezed
¾ cup coconut milk
¾ cup erythritol
1 cup desiccated coconut

1 teaspoon baking powder
2 eggs, beaten
2 tablespoons coconut oil
A dash of salt

Instructions:

1   Preheat the Innsky air fryer for 5 minutes. In a mixing bowl, combine all ingredients. Use a hand mixer to mix everything. Pour into a baking dish that will fit in the air fryer.

2   Bake for 25 minutes at 350°F or until a toothpick inserted in the middle comes out clean

Bread Pudding with Cranberry
PREP: 5 MINUTES • COOK TIME: 45 MINUTES • TOTAL: 50 MINUTES • *SERVES: 4*

Ingredients

1-1/2 cups milk
2-1/2 eggs
1/2 cup cranberries1 teaspoon butter
1/4 cup and 2 tablespoons white sugar
1/4 cup golden raisins
1/8 teaspoon ground cinnamon

3/4 cup heavy whipping cream
3/4 teaspoon lemon zest
3/4 teaspoon kosher salt
3/4 French baguettes, cut into 2-inch slices
3/8 vanilla bean, split and seeds scraped away

Instructions:

1   Lightly grease baking pan of air fryer with cooking spray. Spread baguette slices, cranberries, and raisins. In blender, blend well vanilla bean, cinnamon, salt, lemon zest, eggs, sugar, and cream. Pour over baguette slices. Let it soak for an hour. Cover pan with foil.

2   For 35 minutes, cook on 330°F.Let it rest for 10 minutes. Serve and enjoy.

Per Serving: Calories: 581; Fat:23.8g; Protein:15.8g; Sugar:7g

Spiced Pears With Honey-Lemon Ricotta
PREP: 10 MINUTES • COOK TIME: 8 MINUTES • TOTAL: 18 MINUTES • *SERVES: 4*

Ingredients

2 large Bartlett pears
3 tablespoons butter, melted
3 tablespoons brown sugar
½ teaspoon ground ginger
¼ teaspoon ground cardamom

½ cup whole-milk ricotta cheese
1 tablespoon honey, plus additional for drizzling
1 teaspoon pure almond extract
1 teaspoon pure lemon extract

Instructions:

1   Peel each pear and cut in half lengthwise. Use a melon baller to scoop out the core. Place the pear halves in a medium bowl, add the melted butter, and toss. Add the brown sugar, ginger, and cardamom; toss to coat. Place the pear halves, cut side down, in the air fryer basket.

2   Set the Innsky air fryer to 375°F for 8 to 10 minutes, or until the pears are lightly browned and tender, but not mushy. Meanwhile, in a medium bowl, combine the ricotta, honey, and almond and lemon extracts. Beat with an electric mixer on medium speed until the mixture is light and fluffy, about 1 minute. To serve, divide the ricotta mixture among four small shallow bowls. Place a pear half, cut side up, on top of the cheese. Drizzle with additional honey and serve.

Black and White Brownies
PREP: 10 MINUTES • COOK TIME: 20 MINUTES • TOTAL: 30 MINUTES • *SERVES: 8*

Ingredients

1 egg
¼ cup brown sugar
2 tablespoons white sugar
2 tablespoons safflower oil
1 teaspoon vanilla

¼ cup cocoa powder
⅓ cup all-purpose flour
¼ cup white chocolate chips
Nonstick baking spray with flour

Instructions:

1   In a medium bowl, beat the egg with the brown sugar and white sugar. Beat in the oil and vanilla. Add the cocoa powder and flour, and stir just until combined. Fold in the white chocolate chips. Spray a 6-by-6-by-2-inch baking pan with nonstick spray. Spoon the brownie batter into the pan.

2   Bake for 20 minutes or until the brownies are set when lightly touched with a finger. Let cool for 30 minutes before slicing to serve.

Per Serving: Calories: 81; Fat:4g; Protein:1g; Fiber:1g

Indian Toast & Milk

PREP: 10 MINUTES • COOK TIME: 20 MINUTES • TOTAL: 30 MINUTES • *SERVES: 4*

Ingredients

1 cup sweetened condensed milk
1 cup evaporated milk
1 cup half-and-half
1 teaspoon ground cardamom, plus additional for garnish

1 pinch saffron threads
4 slices white bread
2 to 3 tablespoons ghee or butter, softened
2 tablespoons crushed pistachios, for garnish (optional)

Instructions:

1 In a 6 × 4-inch round heatproof pan, combine the condensed milk, evaporated milk, half-and-half, cardamom, and saffron. Stir until well combined.
Place the pan in the air fryer basket.

2 Set the Innsky air fryer to 350°F for 15 minutes, stirring halfway through the cooking time. Remove the sweetened milk from the air fryer and set aside.
Cut each slice of bread into two triangles. Brush each side with ghee. Place the bread in the air fryer basket.
Set the Innsky air fryer to 350°F for 5 minutes or until golden brown and toasty.
Remove the bread from the air fryer. Arrange two triangles in each of four wide, shallow bowls. Pour the hot milk mixture on top of the bread and let soak for 30 minutes.
Garnish with pistachios if using, and sprinkle with additional cardamom.

Maple-Pecan Tart With Sea Salt

PREP: 15 MINUTES • COOK TIME: 25 MINUTES • TOTAL: 40 MINUTES • *SERVES: 8*

Ingredients

For the Tart Crust
Vegetable oil spray
⅓ cup (⅔ stick) butter, softened
¼ cup firmly packed brown sugar
1 cup all-purpose flour
¼ teaspoon kosher salt
For the Filling

4 tablespoons (½ stick) butter, diced
½ cup packed brown sugar
¼ cup pure maple syrup
¼ cup whole milk
¼ teaspoon pure vanilla extract
1½ cups finely chopped pecans
¼ teaspoon flaked sea salt

Instructions:

1 For the crust: Line a 7 × 3-inch round heatproof pan with foil, leaving a couple of inches of overhang. Spray the foil with vegetable oil spray.
In a medium bowl, combine the butter and brown sugar. Beat with an electric mixer on medium-low speed until light and fluffy. Add the flour and kosher salt and beat until the ingredients are well blended. Transfer the mixture (it will be crumbly) to the prepared pan. Press it evenly into the bottom of the pan.
Place the pan in the air fryer basket.

2 Set the Innsky air fryer to 350°F for 13 minutes. When the crust has 5 minutes left to cook, start the filling. For the filling: In a medium saucepan, combine the butter, brown sugar, maple syrup, and milk. Bring to a simmer, stirring occasionally. When it begins simmering, cook for 1 minute. Remove from the heat and stir in the vanilla and pecans. Carefully pour the filling evenly over the crust, gently spreading with a rubber spatula so the nuts and liquid are evenly distributed. Set the Innsky air fryer to 350°F for 12 minutes, or until mixture is bubbling. (The center should still be slightly jiggly—it will thicken as it cools.) Remove the pan from the air fryer and sprinkle the tart with the sea salt. Cool completely on a wire rack until room temperature.
Transfer the pan to the refrigerator to chill. When cold (the tart will be easier to cut), use the foil overhang to remove the tart from the pan and cut into 8 wedges. Serve at room temperature.

Baked Apple

Ingredients

¼ C. water
¼ tsp. nutmeg
¼ tsp. cinnamon
1 ½ tsp. melted ghee

2 tbsp. raisins
2 tbsp. chopped walnuts
1 medium apple

Instructions:

1   Preheat your air fryer to 350 degrees.
    Slice apple in half and discard some of the flesh from the center.
    Place into frying pan. Mix remaining ingredients together except water. Spoon mixture to the middle of apple halves. Pour water over filled apples.

2   Place pan with apple halves into the Air fryer, bake 20 minutes.

Peanut Butter-Honey-Banana Toast

Ingredients

2 tablespoons butter, softened
4 slices white bread
4 tablespoons peanut butter

2 bananas, peeled and thinly sliced
4 tablespoons honey
1 teaspoon ground cinnamon

Instructions:

1   Spread butter on one side of each slice of bread, then peanut butter on the other side. Arrange the banana slices on top of the peanut butter sides of each slice (about 9 slices per toast). Drizzle honey on top of the banana and sprinkle with cinnamon. Cut each slice in half lengthwise so that it will better fit into the air fryer basket. Arrange two pieces of bread, butter sides down, in the air fryer basket.

2   Set the Innsky air fryer to 375°F for 5 minutes. Then Set the Innsky air fryer to 400°F for an additional 4 minutes, or until the bananas have started to brown. Repeat with remaining slices.

Coffee And Blueberry Cake

Ingredients

1 cup white sugar
1 egg
1/2 cup butter, softened
1/2 cup fresh or frozen blueberries
1/2 cup sour cream
1/2 teaspoon baking powder
1/2 teaspoon ground cinnamon

1/2 teaspoon vanilla extract
1/4 cup brown sugar
1/4 cup chopped pecans
1/8 teaspoon salt
1-1/2 teaspoons confectioners' sugar for dusting
3/4 cup and 1 tablespoon all-purpose flour

Instructions:

1   In a small bowl, whisk well pecans, cinnamon, and brown sugar. In a blender, blend well all wet Ingredients. Add dry Ingredients except for confectioner's sugar and blueberries. Blend well until smooth and creamy.
    Lightly grease baking pan of air fryer with cooking spray. Pour half of batter in pan. Sprinkle half of pecan mixture on top. Pour the remaining batter. And then topped with remaining pecan mixture. Cover pan with foil.

2   For 35 minutes, cook on 330°F. Serve and enjoy with a dusting of confectioner's sugar.

Bananas Foster
PREP: 5 MINUTES • COOK TIME: 7 MINUTES • TOTAL: 12 MINUTES • *SERVES: 2*

Ingredients

1 tablespoon unsalted butter

2 teaspoons dark brown sugar

1 banana, peeled and halved lengthwise and then crosswise

2 tablespoons chopped pecans

⅛ teaspoon ground cinnamon

2 tablespoons light rum

Vanilla ice cream, for serving

Instructions:

1   In a 6 × 3-inch round heatproof pan, combine the butter and brown sugar. Place the pan in the air fryer basket.

2   Set the Innsky air fryer to 350°F for 2 minutes, or until the butter and sugar are melted. Swirl to combine. Add the banana pieces and pecans, turning the bananas to coat.
   Set the Innsky air fryer to 350°F for 5 minutes, turning the banana pieces halfway through the cooking time. Sprinkle with the cinnamon. Remove the pan from the air fryer and place on an unlit stovetop for safety. Add the rum to the pan, swirling to combine it with the butter mixture. Carefully light the sauce with a long-reach lighter. Spoon the flaming sauce over the banana pieces until the flames die out. Serve over ice cream.

Pick-Your-Fruit Hand Pies
PREP: 25 MINUTES • COOK TIME: 8 MINUTES • TOTAL: 32 MINUTES • *SERVES: 2*

Ingredients

For the Pastry

1½ cups all-purpose flour

½ teaspoon kosher salt

¼ cup shortening

¼ cup ([½ stick]) butter, cut up

¼ to ⅓ cup cold water

For the Pies

All-purpose flour

Fruit Filling

1 large egg

1 tablespoon water

1 teaspoon coarse sugar

Instructions:

1   Preheat the Innsky air fryer to 325°F. For the pastry: In a medium bowl, stir together the flour and salt. Using a pastry blender, cut in the shortening and butter until the pieces are pea-size. Sprinkle 1 tablespoon cold water over part of the flour mixture. Toss with a fork. Move the moistened pastry to the side of the bowl. Repeat with remaining flour, using 1 tablespoon of the water at a time, until everything is moist. Gather the flour mixture and knead gently only as much time as it takes to come together in a ball. For the pies: On a lightly floured surface, slightly flatten the pastry, then roll from the center to the edge into a 13-inch circle. Place a 6-inch round baking pan on the pastry near one edge. Using a small, sharp knife, cut out a circle of pastry around the pan. Repeat to make two circles. Discard the dough scraps. Place half of the fruit filling on half of one pastry circle, leaving a ¼-inch border. Brush the bare edge with water. Fold the empty half of the pastry over the filling. Using a fork, press around the edge of the pastry to seal it. Poke the top in a few places with a fork. Repeat with remaining filling and pastry. In a small bowl, beat together the egg and water. Brush over the tops of the pies and sprinkle with the coarse sugar. Place the pies in the air fryer basket.

2   Set the Innsky air fryer to 325°F for 8 minutes, or until the pies are golden brown.
   Cool the pies on a wire rack for at least 20 minutes before serving.

Cinnamon Sugar Roasted Chickpeas
PREP: 5 MINUTES • COOK TIME: 10 MINUTES • TOTAL: 15 MINUTES • *SERVES: 2*

Ingredients

1 tbsp. sweetener

1 tbsp. cinnamon

1 C. chickpeas

Instructions:

1   Preheat air fryer to 390 degrees.
    Rinse and drain chickpeas.
    Mix all ingredients together and add to air fryer.

2   Cook 10 minutes.

Per Serving: Calories: 111; Fat:19g; Protein:16g; Sugar:5g

Cherry-Choco Bars
PREP: 5 MINUTES • COOK TIME: 15 MINUTES • TOTAL: 20 MINUTES • *SERVES: 8*

Ingredients

¼ teaspoon salt

½ cup almonds, sliced

½ cup chia seeds

½ cup dark chocolate, chopped

½ cup dried cherries, chopped

½ cup prunes, pureed

½ cup quinoa, cooked

¾ cup almond butter

1/3 cup honey

2 cups old-fashioned oats

2 tablespoon coconut oil

Instructions:

1   Preheat the Innsky air fryer to 375°F. In a mixing bowl, combine the oats, quinoa, chia seeds, almond, cherries, and chocolate. In a saucepan, heat the almond butter, honey, and coconut oil. Pour the butter mixture over the dry mixture. Add salt and prunes. Mix until well combined. Pour over a baking dish that can fit inside the air fryer.

2   Cook for 15 minutes. Let it cool for an hour before slicing into bars.

Pumpkin-Spice Bread Pudding With Maple-Cream Sauce
PREP: 15 MINUTES • COOK TIME: 35 MINUTES • TOTAL: 50 MINUTES • *SERVES: 6*

Ingredients

For the Bread Pudding

¾ cup heavy whipping cream

½ cup canned pumpkin

⅓ cup whole milk

⅓ cup sugar

1 large egg plus 1 yolk

½ teaspoon pumpkin pie spice

⅛ teaspoon kosher salt

4 cups 1-inch cubed day-old baguette or crusty country bread

4 tablespoons (½ stick) unsalted butter, melted

For the Sauce

⅓ cup pure maple syrup

1 tablespoon unsalted butter

½ cup heavy whipping cream

½ teaspoon pure vanilla extract

Instructions:

1   For the bread pudding: In a medium bowl, combine the cream, pumpkin, milk, sugar, egg and yolk, pumpkin pie spice, and salt. Whisk until well combined. In a large bowl, toss the bread cubes with the melted butter. Add the pumpkin mixture and gently toss until the ingredients are well combined. Transfer the mixture to an ungreased 6 × 3-inch heatproof pan. Place the pan in the air fryer basket.

2   Set the fryer to 350°F for 35 minutes, or until custard is set in the middle.
    Meanwhile, for the sauce: In a small saucepan, combine the syrup and butter. Heat over medium heat, stirring, until the butter melts. Stir in the cream and simmer, stirring often, until the sauce has thickened, about 15 minutes. Stir in the vanilla. Remove the pudding from the air fryer, serving with the warm sauce.

Cinnamon Fried Bananas

PREP: 5 MINUTES • COOK TIME: 10 MINUTES • TOTAL: 15 MINUTES • *SERVES: 2-3*

Ingredients

1 C. panko breadcrumbs
3 tbsp. cinnamon
½ C. almond flour

3 egg whites
8 ripe bananas
3 tbsp. vegan coconut oil

Instructions:

1   Heat coconut oil and add breadcrumbs. Mix around 2-3 minutes until golden. Pour into bowl.   Peel and cut bananas in half. Roll each bananas half into flour, eggs, and crumb mixture.
2   Place into the Air fryer. Cook 10 minutes at 280 degrees. A great addition to a healthy banana split!

Per Serving: Calories: 219; Fat:10g; Protein:3g; Sugar:5g

Spiced Apple Cake

PREP: 15 MINUTES • COOK TIME: 30 MINUTES • TOTAL: 45 MINUTES • *SERVES: 6*

Ingredients

Vegetable oil
2 cups diced peeled Gala apples (about 2 apples)
1 tablespoon fresh lemon juice
¼ cup (½ stick) unsalted butter, softened
⅓ cup granulated sugar
2 large eggs
1¼ cups unbleached all-purpose flour
1½ teaspoons baking powder

1 tablespoon apple pie spice
½ teaspoon ground ginger
¼ teaspoon ground cardamom
¼ teaspoon ground nutmeg
½ teaspoon kosher salt
¼ cup whole milk
Confectioners' sugar, for dusting

Instructions:

1   Grease a 3-cup Bundt pan with oil; set aside.

In a medium bowl, toss the apples with the lemon juice until well coated; set aside.

In a large bowl, combine the butter and sugar. Beat with an electric hand mixer on medium speed until the sugar has dissolved. Add the eggs and beat until fluffy. Add the flour, baking powder, apple pie spice, ginger, cardamom, nutmeg, salt, and milk. Mix until the batter is thick but pourable.

Pour the batter into the prepared pan. Top batter evenly with the apple mixture. Place the pan in the air fryer basket.

2   Set the Innsky air fryer to 350°F for 30 minutes, or until a toothpick inserted in the center of the cake comes out clean. Close the air fryer and let the cake rest for 10 minutes. Turn the cake out onto a wire rack and cool completely.

Right before serving, dust the cake with confectioners' sugar.

HERITAGE OF FOOD: A FAMILY GATHERING

To survive, we need to eat. As a result, food has turned into a symbol of loving, nurturing and sharing with one another. Recording, collecting, sharing and remembering the recipes that have been passed to you by your family is a great way to immortalize and honor your family. It is these traditions that carve out your individual personality. You will not just be honoring your family tradition by cooking these recipes, but they will also inspire you to create your own variations, which you can then pass on to your children's.

The recipes are just passed on to everyone, and nobody actually possesses them. I too love sharing recipes. The collection is vibrant and rich as a number of home cooks have offered their inputs to ensure that all of us can cook delicious meals at our home. I am thankful to each one of you who has contributed to this book and has allowed their traditions to pass on and grow with others. You guys are wonderful!

I am also thankful to the cooks who have evaluated all these recipes. You're, as well as, the comments that came from your family members and friends were invaluable.

If you have the time and inclination, please consider leaving a short review wherever you can, we would love to learn more about your opinion.

https://www.amazon.com/review/review-your-purchases/

ABOUT THE AUTHOR

Vanessa is a New York-based food writer, experienced chef. She loves sharing Easy, Delicious and Healthy recipes, especially the delicious and healthy meals that can be prepared using her Air fryer. Vanessa is a passionate advocate for the health benefits of a low-carb lifestyle. When she's not cooking, Vanessa enjoys spending time with her husband and her kids, gardening and traveling.

Made in the USA
Monee, IL
19 December 2021

85432757R00125